BIRDS WE LOVE TO TALK ABOUT

By:

V. Harvey Mortensen

Table of Contents

BIRDS WE LOVE TO TALK ABOUT

By: V. Harvey Mortensen

My fascination and introduction to bird watching came very early in life while just a small boy exploring and playing in a nearby park. The park was rustic with mostly native trees, flora, and fauna. A part of the beauty of this park was the Smithfield Creek that flowed out of the Wasatch Mountains and through the edge of this fantastic playground near my home. This natural wonderland was where my friends and I played *Hide and Seek*, *Steal the Flag,* and acted out wondrous scenes from action movies of those days, like *Tarzan*, *Bomba the Jungle Boy,* or *Jungle Jim*.

It was on a day while alone that my make-believe adventures had led me near this beautiful, wonderful mountain stream. Whatever dream of adventure I was portraying had caused me to approach this setting, as though stalking someone or perhaps a wild animal. My caution caused care in each step. Not a pebble moved nor a stick or twig snapped. My body was kept low and hidden from this imaginary foe that was in my hot, but cautious, pursuit.

Now properly located in a natural blind, my first peek at this imagined enemy was ventured. To my surprise, there on a shelf of rocks arranged to guard against erosion of the creek bank was a small black bird which was bouncing or dancing. My fascination with this creature, his exuberant zest for life, his song, his dance, his quick flight here and there around his mountain stream home, captured all of my attention.

I observed its performance and its intriguing deep knee-bend dance for a long time. This bird acted as though he were in an empty forest theater, had the stage, and was engrossed in a performance that the viewers or audience would desire. His song was equally full of self-enjoyment, like his dance, a rendition truly given to express love of life and delight in the comfort of his home of attractive natural surroundings. From time to time he would leave the prominent point or stage where performance was required and go to work.

Work consisted of flying to the edge of the stream, wading in with his long legs and then dousing his beak, head, and neck to search under the water for rock-rollers or other aquatic insects for food. After the find, it would quickly fly up stream very near the water to its nest where mate or young or both awaited for the tasty morsels.

When home that evening, in my childish way, I told my mother about this small, black bird and asked her what its name was. My description was poor. To describe (even to my mother who loved nature and especially birds) the joy, affection, and delight of this youthful bird watching experience was not easily performed. Many questions were asked and I did my best to answer, describe, and identify this bird to her. The then unfamiliar terms we use to distinguish birds were not part of my youthful vocabulary.

"Let's go and I will show you, ok?" Mother was cooperative, but our trip was in vain. Many times when one recalls a sighting, the second visit is quite unsuccessful. Yet the previous viewing always causes hope that we could find him right there again. That visualized picture in our minds seemed always to cause anticipation when returning to the spot of a sighting. It took many trips back and many times of coaxing Mom to come before she was blessed with a viewing, which conveyed to her how my interest had been so completely captured.

My mother did not know this bird, although she was born and raised near this Rocky Mountain stream. This unique small black bird was a new discovery to her, as well. A few days later mother came home from work with a huge, expensive book about birds and together we found this bird, "The American Dipper". This was my first introduction to bird watching.

My vocational career was with The Boy Scouts of America, which has provided many opportunities to participate in the study of nature. Bird watching has always been a cherished part of this study. A little extra time was customarily taken, as business took me to and from scout camps.

My scouting career has also provided opportunities to see much of this planet's most beautiful outdoor world. From Hong Kong to Florida, from British Columbia to Southern California, it has been my pleasure to view and love God's creatures of the air. It has been my opportunity to attend many types of meetings, activities, and events where the last words spoken were mine to give. This last short speech in scouting circles is called "The Scouter`s Minute". It has been my practice to use birds' talents, features, aptitudes, character, and niches to suggest some type of moral improvement in our lives, as well as to teach others to love birds, and desire to watch them. To help others to know birds by name and learn their unique habits and flights is also my joy.

Ten years of our family's life was spent in Anaheim, California. My mother and father came down to see us and their grandchildren. While there, mother pointed out to us that a mockingbird was singing for us each morning. She delighted so much in the many refrains of the Mockingbird songs. This caused her to repeat many times, "Oh how I wish we had a mockingbird back home". She enjoyed that bird so much and we had not noticed it was even there. She blessed our lives by teaching us to know and enjoy something we already had.

On one trip from Utah into Wyoming traveling on I-80 east, my council president and I were talking about my love for birds. The freeway goes east near the Weber River in the Rockies. I said to Dave, "Look up there along the road. Those dark spots in those trees are Bald Eagles." He responded, "You're kidding." When we were close I pulled the car off the road, reached under my seat and provided him with some binoculars. After feasting on this scene of three bald eagles for a few minutes he said, "I never before have seen our nation's bird in the wild." Winter brings the Bald Eagles south to fish the not yet frozen waters. Bald Eagle watching in winter can be enjoyed from late November through the middle to end of February in the Rockies (Utah, Idaho and Wyoming).

One of our good friends, Cheryl, insisted that she wasn't going to get trapped into this silly game we called bird watching. As we traveled with her many times we would choose a road which provided many wondrous viewings. She worked hard at her commitment but we would still take time to pull off the road and use the binoculars and enjoy our sightings. Of course, we would offer optical support to her. A spare pair of binoculars is always necessary for these special times.

Not many weeks later we passed through this same special area again with an additional new passenger in the car. To our surprise, Cheryl was now the teacher and she was pointing out where to look and how to use the binoculars. The sighting was a Canadian goose, gander, and seven little goslings in the practice of family school. They were waddling from the bank into the wondrous pond of water for a swim; goose first, seven goslings, followed by the gander.

We started to tease Cheryl about having succumbed to this delightful hobby. She defended, "Well, all of the birds don't get me, but those little goslings are so cute and proper under Mother and Father's care".

My great pleasure is to make people aware of the wonders of nature right around them. Then, almost immediately, I find delight in the opportunity to see them become newly converted bird watchers and out showing and convincing others to enjoy these fascinating creatures that have the gift of flight.

My prayer is that this book will improve or cause you to find moral strength through the study of birds. We spend so much time and effort in the commerce of life. Please, from time to time, take a moment to enjoy these amazing feathered friends of the air. We often suggest, "Hey stop a moment and smell the roses." Well let me suggest that life is better when birding; so why not find a moment to take a look. There is really much that can be learned from sighting and then talking about birds.

Harvey and Patsy Mortensen in Florida, 1991

TANAGER, WESTERN

The vigilant explorers of the Lewis and Clark expedition logged the discovery of this attractive bird. This tanager was originally named the Louisiana Tanager after the Louisiana territory; however, it only intermittently visits the state of Louisiana. Therefore, it became known as the Western Tanager. It's about 6½ to 7¼" long. The bill is pale, conical, blunt and stout. The wings are long. The pointed tail is notched and shorter than the wings.

This species is one of North America's most colorful. The female has an olive-yellow face and two thin wing bars. Its upper is yellow-green and its lower is white. It has yellow under parts with gray black wings and tail. The breeding male has a red face, red to orange to yellow forehead-chin-nape, yellow under parts, and black back, wings, and tail. It has two wing bars, the upper is wide and yellow and the lower is narrow and white.

This brightly colored bird is one sought to be glimpsed during migration and can be occasionally seen at fruit-rich berry trees. This songbird migrates at night and does so silently and is only seen briefly in daylight. When trying to sight this tanager, we will find that this bird moves slowly, and not often, living most of its time high in the canopy. They breed in Alaska, southwest Canada, Rocky Mountains, Sierra Nevada, and Montana, but occasionally visit many other states. They winter in Mexico and Central America.

Camp Loll lies west of Flag Ranch. Flag Ranch is halfway between Yellowstone National Park and the Grand Teton National Park. Scout and Scouter find delight in describing sightings of this brightly colored bird at this attractive camp.

Tanagers like insects, fruits, and some flowers. They feed their young primarily on insects, which they catch while perched or in darting short flights and quickly return to their chosen balance. They love cherries but swallow the fruit whole and limit themselves to smaller morsels, such as choke cherry.

The nest is a shallow cup of twigs, mosses, and grasses built in a conifer tree in June or early July where three to five eggs (pale blue and spotted with brown) are laid.

The Western Tanager sounds like a raspier American Robin. It delights in singing and will stay at a perch for quite a while to render its call, thus allowing us birders to find and experience what we call "a sighting". The detection or incident of sighting of color, dexterity, or song is tricky. The call is a vigorous slurred *pet-er-ick* or *pit-i-tuck* refrain of two or three syllables. Look towards the sound to sight.

Grandmother knitted, crocheted, and embroidered many beautiful things, works of pure art. A few of these delicate, lovely items were seen or used, but most were carefully tucked away in the cedar chest. My sister took piano lessons and became very accomplished but would never share her skill with others. Not at any time would she play at school, at church, or for company at our home. She enjoyed her art and loved many of the works she had mastered. At times when she was enraptured while rendering a composition it was possible to sneak into the room and enjoy her artistry. The Western Tanager is much like these two women in my life. When they are finding great pleasure from their song, they can be found, heard, and sighted for a few minutes of pleasure but usually not for long. What a desire it becomes to witness or experience this colorful tanager.

JAY, PINYON

This bird could be described as a small blue-gray crow (just larger than a robin) about 9 to 10½" long. The bill is long, slender, ridged, and spiked. Distinct bristle feathers are near the base of the mandibles' corners. The head is not crested. The tail is almost square and much shorter than the wings.

The intensity of the blue and gray varies greatly from region to region. This gives to some the delight of watching blue birds but to others the feeling of spying on crows. It is quite bright blue-gray over all, but the head, shoulders, wings, and tail are somewhat darker blue-gray. The back of the neck is sometimes purplish-blue, and the throat is streaked with faint white. The posterior under-parts are a paler gray-blue. Juveniles are gray.

Ornithologists refer to this bird as a very social bird found in flocks of nearly 100 to 250 which nest in loose colonies; but this bird watcher sees Pinyon Jays in pairs in all seasons. They are like crows in rapid direct flight and other customs. These birds roost when not breeding and find roosting sights up to three miles away from feeding areas. The cycle of food it prefers causes a distinctly different breeding season than other American birds.

It is an enchantment to watch them flirt amidst the outstanding fall colors. They become so involved with each other that they know not the bird watcher exists. Courtship occurs in October and November; breeding and nesting take place in late January and early February. When observed walking on the ground, their actions are truly crow-like.

Pinyon Jays like forests of juniper, mixed pines, and oaks in the western United States. They breed from Montana on the north to Arizona and New Mexico on the south and from inland California and Oregon on the west to western South Dakota, Colorado, and the panhandle of Texas on the east.

Their favorite food is pinyon pine nuts (derivation of their name), but they eat all types of pine nuts, oak acorn, and juniper berries. They gather and store these hard fruits to be used in winter and spring. They scavenge like crows and eat berries, worms, insects, grasshoppers, eggs, and the young of other birds.

The nest is a somewhat bulky deep cup of twigs and grasses. They lay from three to five blue-white or green-white thickly and evenly speckled eggs. The female is enabled to stay on the nest and keep the eggs warm even in very cold weather; for the male comes and feeds her from their cache.

Their call notes are quite varied, from harsh cackling or jay-like sounds to *mew* or *quail* notes. When in flight, a *cah-ah*, but when parched, a *queh-quah-queh* song can be heard.

When autumn displays its brilliant colors of red and gold, it seems to suggest to the Pinyon Jay that it is also on display. This is when these enjoyable birds begin to pair and work together. At this time of year, if you see a Pinyon Jay, just wait a minute and you will become aware that there are two, and that they work together. They start their pairing by passing food to one another as they seek the plentiful fruits of the harvest season and prepare to nest, breed, and brood before spring comes.

No matter whether summer, winter, spring, or fall, it creates a good feeling in us all to get a good head start on things; so let's get expeditious, like a pair of Pinyon Jays!

HERON, BLACK-CROWNED NIGHT

From the tip of its pointed, short and stout bill, to the tail of this medium-sized heron are 23 to 28" and have a wing span of 44". In flight, the feet project beyond the tail while other heron show legs and feet. Atop its crowned head is found occipital plumes which stream across its sturdy back. This short-necked, hunched postured bird seems always to be in a lurking pose, whether dosing in daylight or hunting at night. Slumber time reveals the Black-Crowned Heron to be a most gregarious nesting bird.

These communities are created usually in isolated thickets, and their population may be as large as several hundred pairs. As young hatch, each nest is increased by four or five more new young.

Usually these night herons hunt alone in favorite sites likely several miles from where chicks anxiously await morsels to eat. The work to sustain new life is constant and burdensome. A colony of adult and immature Black-Crowns blend so well into the foliage of marshes that it is impossible to see that the young outnumber the adults, until flight is aroused by all.

Adult (sexes are similar) have blue-gray bills touched with yellow; large red eyes, blue-black cap and back, pale gray wings, white under-parts, and yellow legs and feet. When breeding, legs and feet turn bright pink, and their plumes become extended. Immature upper-parts are light gray-brown and feather tips have large white spots. Under-parts are streaked dull-white and light-brown. Adult color is acquired in three years. During their first spring, under-parts are paler (breast light tan, belly tan-white), upper-parts are darker (crown and back dark brown), wings are medium to light brown. By the second spring, they have almost obtained adult color (belly white and wings light brown-gray). This heron seeks fresh water rivers, afforested swamps, and marshes that are brackish for breeding, and do so throughout most of the United States from the Rocky Mountains to the Pacific Ocean in the west. In the east, from south-central Canada to south-east Canada on the north; and from the Gulf of Mexico to the Florida Keys in the south. They also are multiplying in Eurasia and Africa.

Well after egrets, bitterns, and other herons have vacated fishing waters to prepare for evening sleep, the night heron continue to search for the nutrients they need. They hunt briskly, holding their heads low with necks curled, all ready for a sudden strike at fish or frogs.

The nest is a rather haphazard pile of sticks and branches, lined with twigs among reeds in marshes or up high in a tree where 2 to 6 pale bluish-green eggs are laid. The colonies of young leave their nests in about 40 to 48 days after hatching.

The Black-Crowned Night Herons are heard at dusk or night and the yelp or bark is given before stalking, as well as in quick (flapping wing) flight. The harsh *"squawk," "quawk or kwak"* gives rise in some regions to nickname this heron *"The Night Squawker"*; but to the true birder, it is a true delight to listen to!

This bird teaches us that night-life can indeed be good; for it lowly exclaims it is fun to be up late. Yes - food, drinks, and company, what a pleasant anticipation! Morning people rise early, clean and fuss all day - "Oh, dear." "Hooray!" What a nice premise to find in nature that the Black-Crowned Night Heron needs a night life! So go have fun, enjoy yourselves, but don't stay out to late, ok?

STILT, BLACK-NECKED

This is a medium-sized shore bird or wader. The legs are as long as (from head to tail) 13 to 14" with a thin elongated bill. In flight the legs extend causing a silhouette image of nearly 2 feet. For bird watchers in southern and western wetlands, the stilt is the most elegant and easily recognizable as what one might call a "skinny bird". Stilts have the longest legs in reference to their bodies of any bird.

As a Weight Watcher, I presented a re-print of this "skinny bird" to my exuberant instructor, Mary Ann. She claims to view this Black-Necked Stilt picture for inspiration to keep the weight off as she prepares for the challenges of each day.

Their glossy black back and upper wings contrast sharply with pure white under-parts. It has a needle-like black bill with a white spot above and behind each reddish-brown eye. Its long legs are a red-pinkish color. Black continues up the back of the neck to the top of the head. The chin through the breast and belly are white. The tail is an ash-gray color. In the female, the black is more dingy. The immature upper parts are brown.

When the stilt wades in the salty shallow waters, he lifts his foot with his very pronounced long legs, like each foot got stuck in the mud and is quite hard to pull out; yet he steps forward with the daintiness and elegance of a fine butler in black tie and tails. In flight, the distinct contrasting colors seem even more defined. The long needle-pointed bill stretching straight out front and the lengthy legs dangling behind gives an exaggerated streamlined appearance. It flocks with other stilts. They defend their territories by screeching their song vigorously and flying like buzz bombers at the intruders. In order to cool and care for its mate, its eggs, and eventual chicks, the stilt flies from water source to nest during incubation, moistening ventral feathers and using evaporation to help cool chicks.

They live in open seashores, lake shores, and shallow fresh water. They breed from southern Oregon to northern Utah, from Nebraska to Texas, and Delaware to southern Peru.

The nest is a depression in the sand, open ground, where they lay three to five buff colored eggs with spots or blotches of dark brown. Incubation takes 22 to 26 days by both sexes, and the young leave the nest four to five weeks after hatching. Rising water may cause the stilt pair to construct a well-built floating platform for the eggs.

Stilts eat aquatic insects, snails, beetles, grasshoppers, and black bugs.

They call a *Kek-Kek-Kek* or a yelping *Yip-Yip-Yip* sound.

Life is full of brackish waters and other messy things. We should approach these difficulties with care. Each step forward should be made with care, precision, and delicate vigilance. Always looking good, white shirt, dark coat, but aggressively working for a fine future. When necessary, we should join with others to sound the cry of alert and drive off those elements which would destroy good life. We should stand a little taller and protect what is right. We ought to step through difficult times like the Black-Necked Stilt, keeping ourselves distanced from the muck in life as we lengthen our stride towards excellence.

HAWK, SWAINSON'S (Dark Phase)

The Swainson's Hawk is 19 to 22" long and has a wing span of nearly 52". This large hawk when perched shows long and pointed wings, which sometimes extend past its tail. The Swainson's Hawk is often seen (spring and autumn) in great flocks when migrating north or south. Sexes are similar, but females are larger than males. Each soars effortlessly (with wings held somewhat high, forming a V) on uplift wind with ease.

This hawk is among the world's long-distance champions, flying from Alaska to Argentina, using wing beats which are rather quick and light for a large hawk. These raptors fly by the thousands in early spring over the Rio Grand River from Central America to territories in the United States (Alaska predominately included) and Canada. The flocks become small as, one by one, the birds reach their individual chosen nest areas. These nest sites have been used by their progenitors for many years.

The maturity or immaturity of the Swainson's and the phasing from dark to light cause much variation in plumage color. The <u>light phased adult</u>, has dark tipped yellow mandibles, a white mask (chin and forehead), a lighter brown head and throat, and dark tan dusky bands.

Its under-parts are cream or white and lightly barred and spotted, except for its dark gray flight feathers. The wing linings are very whitish and lag. Its feet are yellow with black claws. The dark-phase adult has all dark brown plumage. The colors on the immature are fragmented or spotted.

They are widely distributed across the western United States and Canada. They nest as far north as Alaska and winter on the plains of Argentina.

They prefer grasslands, plains, prairies, agricultural areas, and deserts. This bird spies for prey from high perches or it hovers

brown back wings. The tail is grayish-brown with nine or

while hunting. The food of the Swainson's Hawk consists of smaller rodents, principally gophers and mice, as well as large insects, occasionally snakes or small birds. From an economic point of view, this falcon is considered to be the most useful to man of all of the hawks.

The nest is a rather flimsy large mass of twigs and grasses on ground, ledge, or in the tallest tree on a horizontal branch. They lay two to four white, sometimes faintly marked with brown, eggs. The incubation period is about 28 days by both sexes.

This raptor is usually silent but occasionally gives a long, whistled *Kieeeer,* quite highly pitched.

This hawk is amazingly different. This is a hawk joins with other Swainson's to migrate in massive flocks for very long distances, is willing to leave the massive group for family, knows from heritage what lands to call home, and has dramatic and delightful plumage variations. Now what can we learn from this bird of prey? Well, if you remain at or near home, never stray away. All you will need is a nice change of dress.

GOLDFINCH, AMERICAN

The American Goldfinch is a small perching bird, 4 to 5" long with a somewhat stocky body. It's a very friendly little bit of brightness and is enjoyed in many backyards by those of us who choose to provide bird seed for them.

The breeding male is bright yellow with black cap worn slightly forward or over its brow. It has a white rump with shallow wing bars, and white edges to its black tail and wings. In late fall, after nesting, the male turns more olive yellow-green, much like the female. The female is olive yellow-green with no black cap. Both have a conical dull orange bill. Winter brings a change in plumage. The male has a yellow-brown head and back. The female has a gray-brown head and back. This minor change in tone doesn't hamper the spirit of these delightful birds.

This is a bird my grandchildren love. Easily attracted to bird feeders, the Goldfinch at close viewing always seems cheerful, especially when the seed mix is to its liking. These birds bring cheer, delight, and songs of hope to grandchildren, as well. Of course, there are smiles of delight by youthful boys and girls as they point out to each other this yellow bird's wonder, care, and dexterity as it peels a sunflower seed right in front of their eyes. It is especially interesting to watch the Goldfinch feed on a thistle because of the detail, care, and accuracy they use to attentively search through and select the prong-type leaves to find seeds. Somehow they extract the meaty kernel (most important part), dropping the shell which, without detailed inspection, appears untouched. This same passion for detail and perfection is displayed in the building of a nest as they carefully construct it so well that it will hold water. They are sometimes called thistle birds. When the rains come, the adults act as shields, a perfect canopy or umbrella as they perch on the nest edge, spreading wings and tail to keep water out of the nest and off the young.

Goldfinches finish their abodes in weedy fields, farmlands, meadows with shrubs, and open tree stands. They are found from coast to coast, from southern Canada to Mexico. They love thistle and sunflower seeds, but many other seeds are relished in their diet. In summer they will add insects, such as larvae, grasshoppers, caterpillars and aphids.

They nest later than other birds, during the time there are plenty of seeds maturing. They lay three to five bluish-white eggs in a nest of grasses, down, and bark. The nest is built in tree forks from 5 to 30 feet above the ground.

The finch's song is a mixture of trills, twittering notes followed by *swees*. In flight it calls, *per-shic-o-ree, per-chic-o-ree,* and seems to enjoy flight and song deep into summer, but when nesting starts the song becomes more businesslike, *tec-o-ree, ba-ba, ba-bee.*

Beauty, color, and fine formal dress should not cause pride or haughtiness. Elegant attire can be used to magnify a cheerful, friendly disposition. Where melody and encouragement are constant and a part of life, fulfillment in homemaking, vocational work, or harvest restraint is then avoided. Instead, the brightness of life along the way is always available. And when autumn comes, Goldfinches mix well in great swarms of flight (like roller coaster rides) always talking, always chirping, helping each member of the flock enjoy the sunset skies.

OWL, BARN

This bird of prey is a medium-sized owl, 14 to 18" long with a wing span up to 44" in length. It has a distinctive heart-shaped face where in is found a whitish-yellow mandible. Legs, feet, and talons are exposed. The bill is almost hidden under soft plumage. When this owl is perched it sits quite upright.

Barn Owls are golden brown above with gray, white, and pearl colored flecks or spots. Under-parts are white, touched with spots of brown-gray. The dish face has a distinctive light-buff color, where the long narrow beak is centered. Eyes are dark purple-brown with some feathers very near the eye. The white concave face is unique. An outline of the love-shaped face is accomplished by a darker buff nap.

This raptor is much more tolerant of men than other owls. The dished face is set just right to cause the earflaps to catch even the slightest sound. Combining this feature with dark night vision eyes causes a very successful hunting creature of the night. It approaches game on soft feathers, creating a silent flight to capture prey. Their destruction of rodents, mainly mice and rats, should make them most desired by farmers and ranchers, though some still do not know their value and seek to kill what they would call a spooky creature.

They are wide-spread throughout the world: North and South America, Canada, Europe, Asia, Africa, and Australia (but not New Zealand). This owl is not migratory, but found in savannah, marshland, farmland, and residential areas. This owl avoids conifer forests and mountain ranges.

Barn Owls, as do other owls, digest all soft parts of their victims but regurgitate the bones and fur. They will devour their own weight in rodents every night and can pick small mammals from the ground and birds from a perch or the air. One observer tells of this owl swallowing nine mice; then after bringing the bones and fur up, (three hours later) downing three more.

They build nests on flat surfaces with little or no comfort in old buildings, like barns and on cliff or in coves, and sometimes in trees where limbs create somewhat flat surfaces. They usually lay four to nine round white eggs.

Their song is a long rasping "*screech*" followed by a "*hissing*" sound which seems to get louder as the sound lengthens out. There isn't a single "*hoot*" from the Barn Owl.

As car headlights catch the white under plumage of this stealthy night hunter, one may be startled. To be suddenly surprised by its "*screech*" on a silent evening may be a fright, but not to those who love to sneak through and around old barns at night. As young boys -Rod, Larry, Mark, and I wished to give credence to the fourteenth century characterization of woe, mischance, or evil in the air attributed to the beautiful Barn Owl. Our steadfast searches (for which only boys would hunt) of fright or horror in the night never came; we were always disappointed! We did find caches of bones and hairballs though; thus, keeping our desire to be terrified alive. We wished for it to come, we sought for it to come, but many nights of our young man's game never resulted in a good scare night! However, midday barnyard fun found us sighting a sleeping Barn Owl up above the hay lofts one day. Either Barn Owls sleep soundly or they have no fear of hoodlum youngsters; for neither feather moved nor head turned. It was as though we boys had never found this owl. What a pleasure the searching of nature is even as a youthful schoolboy! Only youth seek to be frightened but are secretly happy it never came.

JUNCO, DARK-EYED, "Oregon" or "Pink-Sided"

Ornithologists suggest there are twelve or more subspecies in the Dark-eyed Junco family. The above painting was composed from several photographs taken of Juncos while perched in a willow tree just above our back porch. They seemed to be waiting their turn to feed, but ravaging House Finches dominated the bird-seed feeder. This unique, somewhat chunky little sparrow is from 5 to 6 ½" long and has a short bill and long notched tail.

This bird is a hopper, never walks, a ground dweller, and also moves through the lower branches of trees and shrubs. In winter, these fairly tame small birds fly up flashing white outer tail feathers, seemingly saying, "Come fly with me." Juncos mix with many other birds to form great swarms to search for food when frost and snow covers the ground. In the 1930's the Dark-eyed Junco was known as the "Snowbird". Ornithologists then and now differ on the groupings of this dark eyed bird; however, in this narrative we will refer to five groups.

"Dark-eyed" Junco: pink bill, black head, brown back, pinkish sides, white belly and outer tail feathers. "Slate-colored" Junco: pink bill; dark gray head, back, breast, sides, and flank.

"White-winged" Junco: very similar to the "Slate-colored" but has a great deal more white in the tail feathers. "Pink-sided" Junco: broad pink sides and flanks with pale grayish breast, head, and back. "Gray-headed"

Junco: bright reddish-brown back; light gray head, breast, and sides.

Juncos show some of the greatest geographic plumage variations of any North American bird. One sub-species, the Gray-headed, is different in its call, as well, which resembles that of the Yellow-eyed Junco.

Each sub-species roams different territories, but on the whole, they breed from Alaska to Newfoundland and South through Northern States but extending further South through Western Mountain States and Appalachian Mountains. This bird spends spring and summer in openings near the edges of conifer forests and mixed woods. Then in winter finds suburbs in open fields and farmyards where they like to roam with other feathered species near their size. Junco groups breed freely with other cousin groups where ranges overlap.

Juncos have a peculiar niche; for they feed mostly on the ground, eating grass seeds, insects, and some small berries. The birds feed at and under the bird feeder behind our home. The House Finch, Rufous-sided Towhee, House Sparrow, and Red-shafted Flicker will all briskly eat while flipping seeds out of the feeder. The Dark-eyed Juncos eat below on the porch. This action reveals patience, staying power, tolerance, or politeness to me. What a nice little bird.

Their nest is built on the ground usually in a natural depression generally somewhat protected by an embankment, stump, or tall vegetation. The nest is cup-like and made of roots, grasses, moss, and twigs. Typically they lay three to six eggs, which are bluish with dark brown speckles.

Birders can easily find occasions to hear this bird, for they love to sing. Their song is a delightful musical ringing trill at an almost continual pitch, occasionally there is a abrupt change of pitch right in the middle of the song *"Tsip", "Zeet", and "Kew Kew".* The song of the Juncos can be confused with that of the Pine Warblers, whose call is very similar.

Progression through life usually leads to what we might call sublime heights. To attempt to manage others (men and women) from heights of loftiness has many precautions. We are taught, "You must feel the soil," "Where the rubber meets the road," and "Seeing that the water gets to the end of the row". To communicate with a child one must get down on the carpet and play with him or her. Juncos look up from the trenches. Perhaps we should as well?

BLUEBIRD, MOUNTAIN

Bluebirds are thrushes and belong to the same family as the robin. They have a forward flight and hover like hummingbirds. This bird is 6 to 7" long. The Mountain Bluebird perches in a more upright position, hovers more often, and has longer wings, tail, and legs than the Eastern or Western Bluebirds. The adult male is a gorgeous sky-blue colored bird. The wings, tail, and some upper parts are a deeper blue. The eye ring, lower belly, and under coverts are white. The female's head and back are grayish brown with bright sky-blue wings. Sometimes in fresh plumage, an orangish cast to the chest and upper sides is shown. The tail and rump are sky-blue. The belly and under-tail coverts are white. The female also shows an eye ring. Juveniles are erratically spotted white above and randomly spotted dark below.

Like roosters crow to announce the dawn of each day, these lovely bluebirds herald the rising of the sun with their warbling "*phew-lor* "and "*tru-lee-tru-lee-tru-lee".* As the sun begins to gleam and rises into the sky, this morning chatter between bluebirds then ceases until another day. My wife and I have had a nest box up for nearly twenty years, but as yet have never attracted a bluebird. My son, David, and his family have bluebirds, but they don't appreciate them chatter before dawn. My oldest son, Brad, and his family, being early risers love the sweet chatter of bluebirds near their home. These Mountain Bluebirds are quite easy going in their approach to life. This makes them vulnerable to other bird species which show aggression in claiming hole or cavity dwellings. For example, Starlings chase Blue Birds out of their nest boxes and just take over.

The Mountain Bluebird is quite comfortable at elevations over 5,000 feet and has been sighted at heights of 10,000 feet during hot summer months.

Bluebirds seek sheltered places to nest: nooks, crevices, old woodpecker holes, or man-made nest boxes. Mountain Bluebirds breed in sagebrush plains, mountain meadows, and timberland areas throughout North America. They breed from north to east Central America and Manitoba or the Dakotas, south from California, northern Arizona and New Mexico. If food remains plentiful, the bright bluebird becomes a somewhat more occasional sightings against snow covered mountain slopes.

Bluebirds are fruit eaters, like thrushes, robins, and waxwings. They also eat insects off the ground or hawk them out of the air.

Men, through bird watching associations and some bluebird organizations have begun to save the three bluebird species. Nest boxes have been distributed across North America and Canada. The nest box trails have saved this beautiful winged creature. Nest box trails can be found over many mountain miles. Especially in the four states that have named a bluebird as the state bird. Four to eight pale blue eggs are laid that incubated 14 days by both the couple, young fly in 21 days.

These bluebirds seem to symbolize good things. As newlyweds seem anxious each morning to visit and renew the wonders of recreating their new friendship, Mountain Bluebirds with their warbling song each day awaken in us memories of love of life. Yes, as man has organized and given time, money, and effort to provide nest homes for the bluebirds, the bluebird have given back the saver of experiences not long past. Yes, our time on earth better with bluebirds.

HERON, GREAT BLUE

This is the largest of all herons, 50 inches with a wing span of 7 feet. It stands nearly 4 feet tall, displaying a long sharp yellow bill. Its white head is crested with two black plumes which begin above the eyes and extend backwards in sharp points. Its flights are not usually long but are sweeping and delightful as they swoop down, cock their necks back, extend their long legs as they abruptly lift upwards, and then gently splash down in wading waters. The adult's long neck, body, wings, and tail are gray-blue. Its head is white with two black crown plum stripes.

The Blue Heron takes off most heavily on cumbersome wings, but soon appears most beautiful in flight with head seated back on its shoulders, neck poked out under its head and bill with its legs extended and trailing. This heron is most deliberate, cautious not to scare off any potential meal. Every step, every movement is designed with care and in quite slow motion. Its head and neck are always ready to show its skill at catching fish or other prey. After catching a fish, he will sweep it in the water as if to wash it. Then he manipulates the catch into head-first position above his head. The catch is swallowed whole.

The Great Blue Heron lives in marshes, lakes, rivers, and shores from Lower Canada, south to Mexico. It likes tall trees and ledges, but will use bushy areas.

In one bird refuge we frequently visit, high wooden power poles with 15° rising limbs now attached provide Great Blue nesting sites near these safe haven waters. Near where the limbs attach, there are two straps on each side, from pole to limbs making a crude cup in which to start the nest. They somehow build large platforms of sticks for their nests in what might be called a community. The colony can be as large as 30 to over 90 stick homes.

This Big Heron is usually silent, but when it's startled, a distinct "*kraak*" is made.

We had a most unique and entertaining viewing. One day a Great Blue Heron came down from building a nest in a willow tree. Pat and I were using our car as a blind. This bird attentively approached us, one slow step at a time. Then it carefully looked all around, took a few more steps, and surveyed all around again. This action would repeat again and again until, surprisingly, right in front of us.

The bird seemed to have finally found just the right proportioned stick to pick up. It was nest-improving time; for herons had colonized here for years. After returning to the nest, the application began for just where or how this new lumber would help the nest. It was tried here, and then tried there, and finally it was used to shore up the very edge of the nest; but it did not fit well and fell to the ground. We expected this great bird to fly down and retrieve this stick, but instead the stick that didn't work was left and the hunt for new material began again.

Caution in life is good when it helps find what is right; for when young herons begin flying practice, the strength of the nest is important. When stress comes, it is nice to find we have built well. As gangly young practice on nests high above the ground, it sometimes seems that they are impatient. Yet, while they clamor around on nests aloft seemingly about to fall from heights which could maim, the truth is that this roughhouse excitement causes young wings to be extended, exercised, and strengthened. Flight in life comes from practice in youth and natural-given exuberance, which when launched from a sturdy platform of truth and hope, will cause assured and true flight.

GREBE, WESTERN

The Western is the largest grebe in North America. This striking swan postured submarine-like diving and swimming bird has a long neck with a body length of 22 to 29″. The sharp dagger-like bill is about the length of the head. Its short wings are invisible and tail a mere powder puff, legs are short and far back on the body, feet are lobed and used for navigation both in the air and under water. Both sexes look alike, and the Clark Grebe, which could be called a separate species, looks very similar. Ornithologists over the years have and haven't called these two the same species. Each has long slender swan-like necks about as long as their bodies. The Western's black crown extends below the oranges-red eye, while the Clark's white cheeks extend above the eye. The bills do differ; the Western's is pale greenish-yellow or yellow, while the Clark's is orange-yellow. Ornithologists suggest the two rarely interbreed. The head, neck, and body show a strikingly sharp contrast of slate-black above and white below. The wings show white wing strips in flight.

One of the great stunning thrills for birders is to see a pair of grebes perform their courtship water show and exhibition. They glide side by side up-turned across the water with necks bowed in an S-shaped surprisingly ballet-like posture. The performance appears to conclude when each drops to a restful glide in the water. But no, this is only the first act. Now each quickly slips away in a surface dive, disappearing from view, yet to emerge breast to breast each with a morsel of moss or underwater vegetation to share with its partner. They have even been seen with necks entwined in a fond embrace. This act of love or sharing turns into the last act of a beautiful celebration expressed as a surface water dance as pirouettes and twirls are shown. Their posture of affection is extended in their lives as they take their young hatchlings along to protect and nurture them. Although as soon as the young are out of the shell, they can dive right in and swim, yet they are invited aboard their parents' backs. This again provides a unique viewing. What a picture! The young show comfort and delight in being brought along, and the parents are proud to show them off.

They can be sighted west from British Columbia and California, east to Lake Winnipeg and the Dakotas. They spend their summer on lakes and marshes where rushes are found and winter together mostly on the pacific coast.

Grebes pursue food, mostly underwater fish, mollusks, and crustaceans, but will eat surface insects, as well.

Grebes almost lay their eggs in the water, but not quite. The nest is a floating raft of marsh vegetation, where three to six eggs are laid. Both sexes care for the young, but will completely conceal the eggs with moss while going to catch fish.

During a courtship display, the grebe may give "*whistles*" and "*wails*", but the Western normally calls a two note "*kreek-kreek,*" and the Clark a mere "*krrk*".

You can be watching a beautiful Western Grebe and right before your very eyes it will disappear, hardly even a ripple, just gone from view. Have you ever just wanted to get away for a while like that? Where do they go? Well, perhaps they go fishing or just for a sight-seeing swim. Any way, they just get away. What do you think? Do you need some time of your own? It is fun to watch them dive, count the time they're down, and then guess where they will surface again. Time away is good, isn't it?

BLACKBIRD, RED-WING

One of the most sighted birds in North America, the Red-Wing Blackbird, is 7½ to 9½″ long. It has more rounded wings and a stouter bill than other blackbirds. The eyes are dark, differing from Rusty or Brewer's Blackbirds which have whitish eyes. The males are glossy black (under some light the black gives off a greenish-blue gloss) with a bright red or reddish-orange shoulder patch which is fringed with buff-yellow on the lower wingbows. The female is dark brown above, tinged with gray and black. This ladybird has a light or buff eyebrow and is pale underneath with heavy bars.

The Red-Winged Blackbirds have their own pattern to establish mating or nesting colonies. First to arrive at the nesting fields are the maverick birds, the strays, and the vagrants. Why they come first is a question to birders. Perhaps the intention is to observe the male Red-Wings' arrival and just learn from them. The highly plumaged males come next and each establishes or claims territories that perhaps they have held years before. Finally, the females arrive about two weeks after and the frenzy of display and mating begins. Each male desires to prove that he has the most brilliant red and yellow wing patches of any male in the colony. Each male seems to invite into his claimed territory one or more females, each free to build nests, lay eggs, brood, and raise young. The proud, blustery male now spends life defending territory or on constant sentinel duties. Red-Wings fly in flocks in winter, sometimes immense in size, not just flocks of thousands, but as many as millions. However, flocks of this size are not all Red-Wings. In these large flocks are grackles, starlings, gold finches, cowbirds, and other species of blackbirds.

Red-Wings seek fresh water marshes with reeds, cattails, and thickets. They breed from central Alaska, southern Yukon, northern Manitoba, and southern Newfoundland-south to Baja, California, Central America, and the West Indies. It has been suggested that the Red-Wing Blackbird may be the most numerous of all birds found in North America. Three to five blue-green eggs which are heavily marked with what appears as reddish-brown pen blotches near the large end of the egg in a woven cup nest attached to reeds, cattails, or tall grass.

The Red-Wing Blackbird eats seeds, grains, insects, and spiders.

These birds vocalize under what appears to be great effort. They sing and display in an awkward motion appearing almost as if it hurts to call a liquid *"cank-ke-ree"* or *"ok-a-lee"* which trills at the end *"tee-ay"*.

The male shows some dignity, good manners and provider skills as he goes before the female and prepares a way for her. The female shows some good sense by allowing his boastful male pride as he defends their territory in a raspy voice and colorful display of his red and yellow wing patches.

There is beauty and good in playing our supporting roles in life. Is it not good to see a man and wife take care of each other? He takes her by the hand or offers her an arm when they walk together. She speaks softly in her council and assistance to him. Manners of support to one another are not old-fashioned but right in life today, as well. The swamp of the neighborhood *"cank-ke-ree"* and *"ok-a-lee"* brings a feeling of support and unity to the bogs they call home.

Have you ever experienced the tranquility of dusk (softly falls the light of day) near the echoing choir of Red-Winged Blackbirds? Calm, serenity, and harmony all comfort the heart and give us life for another day, don't they.

WAXWING, CEDAR

The Cedar Waxwing is a relatively chunky bird that has a sharp crested head; which is more conspicuous when perched than in flight. They have pointed wings, a short tail, and are 6½ to 8″ long. The plumage is very trim with elegant markings, so delicate or distinct they look as though an artist painted them on. It provides most beautiful sightings in early spring. My mother, Lula, was always pleased as they would feast on crab apple and juniper berries just outside our large corner kitchen windows. For just a short time each year we would eat breakfast, it seemed, with these colorful birds. They have pinkish-brown upper parts and breast with a bold black mask through the eyes. The mask is slightly edged above and below with white. The chin is black. The wings are black-gray. The forward secondary feathers have red, seemingly waxy tips on the feathers and have white edges on the back secondary. Its belly is yellow with white under the tail and black-gray above. The tail feathers are tipped with bright yellow.

A highly gregarious bird, the waxwing flies in flocks of 40 or more at migration times and in winter as they search for food. The swarms are even larger in fall, making the sighting a banquet of great pleasure to the watcher. Waxwings are often viewed in a polite exercise of passing food (a berry or flower petal) from beak to beak, as though all were too well mannered to take the last morsel. Some ornithologists suggest this practice is just a game; but if just for fun, it surely is a courteous and respectful frolic. This behavior pattern seems to be innate in the waxwing. For this performance has not been learned from others. Although waxwings seem never to have a feather out of place, at times when berries are overripe or fermented they have been observed to become quite tipsy. Drunkenness for birds is as dangerous as inebriation for men. They can fall from perches or fail in flight, making them vulnerable to become prey for a cat or rodent.

They breed from coast to coast, from southeast Alaska to central California in the west, and from Newfoundland to Georgia in the east. They are found in open ranges, mixed forests, parks, or gardens where fruit or flowering trees are found. They use no specific migration routes, but instead wander to find food as they travel.

They depend mainly on fruits of native plants, such as elderberries, choke cherries, raspberries, etc. The cedar tree's fruit provides one of their favorite foods. In addition to berries, Cedar Waxwings eat flower petals, sap, and insects. Berries are their main food source in the winter, but in the breeding season and summer, insects become more important. They catch insects in flight like flycatchers.

A nest made of loose twigs and grasses is built 6 to 50′ above the ground in a tree; they lay three to six pale blue or grayish eggs dotted with black.

They give a high sibilant "*see-e-e-e*" as their call. In mating season, the waxwing becomes quite unseen and its call becomes the only means of detecting its presence.

Perhaps we should look to the Cedar Waxwing as our example for attending a fine affair and dining experience. First, we should be well groomed. We should use elegant taste in color, texture and pattern of dress; not a feather out of place. We should come with a cheerful, pleasant attitude, ready not only to enjoy the meal, but being socially proper or totally polite. If one studies and learns names it makes it easy to include everyone in the chatter and amusement. A key to this social pleasure is respectful, honorable behavior or manners. A part of this etiquette truly relates to indulgence; for when fermentations are served, we should be prudent, thus never falling from our perch in embarrassment.

KESTREL, AMERICAN

The American Kestrel is the smallest of the falcons, 8 to 12″ long with a 20 to 23″ wing span, jay sized with a raptor bill. Its hunter eyes are eight times stronger than man's.

The male has blue-gray wings and orange-red ruffs, back and tail. The tail is banded black with white tips. The female has orange-red on ruffs, colored wings and back. Each has distinct dark markings on the neck. They look as though an artist placed four black downward brush strokes on the white neck: one below each eye and one below each ear. It is pale under the wings where spotted patterns produce beautiful wing shadow designs. It is the most colored of all falcons.

This small hawk (sometimes called a Sparrow Hawk) will watch the bird watchers carefully. When perched to survey his domain, the tail gently moves for balance. He can be seen, as he constantly keeps watch of us birders (my son Rod and I). The Sparrow Hawk can be driven by intent to get just a little closer for perhaps a better view. Then, knowing we are watching, he loops forward down the string of power poles. This viewing and escaping from view exercise can repeat again and again as we pursue our study forward. Then all of a sudden, instead of pursuing forward, a large looping flight is flown back onto the line of perches from whence we all had come. Thus, he remains in his chosen territory and avoids our further pursuit.

They are the most commonly sighted of all falcons and can be seen above farmlands, forests, grasslands, and cities. They are found from Alaska and Canada, south to Central America, and from the Atlantic to the Pacific.

They lay three to five cream colored eggs blotched with brown in tree holes, nooks in buildings, and sometimes in old magpie nests. The female's unique larger size perhaps helps when brooding. The male feeds the female while on the nest.

On pointed rapidly beating wings, this small hawk hovers then dives to capture rodents, birds, or insects. These undersized raptors have enormous appetites. Estimates suggest each adult rids the world of nearly 300 rodents annually. Other winged creatures find themselves under the kestrels' sharp talons most often, but mice are their principle meal. The opportunity to get real close to study the kestrel is always there when his dinner is before him. In fact, to

slowly walk up and catch a feather dropping while he dines is possible for a careful birder. They vocalize a sharp *klee-klee* as they soar and glide.

One of the neat things about retirement is the fact that my wife and I get to go out to dinner together quite often - good food, something nice to drink, and conversation with your best friend; that is nice. Now let me ask you, have you ever liked eating out alone? No matter how good the food is, it's not as nice as dining with company. Male or female, the American Kestrel always dines alone and seems to enjoy it. Even under close observation their meal seems not to ever be disturbed. They just seem to really enjoy every morsel. This American Kestrel's appetite finds solitude ok. How strange! He is not like you and me.

CRANE, SANDHILL

This bird has long legs, neck, and bill, a plump body, and is 34 to 38″ long. The wings are broad and span 6 to 6½′ in length. The female, although identical in dress, is smaller than her male partner. Cranes are near the height of a man, but the Hoping Crane is taller than the Sandhill. Their drooping tail feathers from the lower back carry down over the wings and form a bustle. This crane is mostly gray with a patch of bare red skin on its crown. The bill and legs are dark and the chin is white. In spring, Sandhills paint their feathers while preening with rust stained mud which oxidizes and combines with the feather pigment giving them a reddish-brown color. Juveniles have a brown forehead, not red.

Although the Sandhill Crane is similar to the Blue Heron, there are distinctions. Sandhills fly in large noisy flocks at high altitudes with neck and legs outstretched with a distinct uplift in their wing beat; while Blue Herons fly alone with their necks curved back and their heads resting on their shoulders. When wading, Sandhills have a relaxed casual appearance, while Herons have a stiff, cautious hunter look.

When the young colts hatch, they can walk at once, and the parents commence a skillful stratagem of the protection of their young. Colts are immediately shown how to forage for seeds and insects, but when intruded upon, they are quickly hidden and each parent takes a separate role. The female acts as though she were wounded and begins to stumble or limp off, leading the intruder away in the process. The male takes a different direction away from the colts, placing himself ready to get in between its offspring and the enemy. His bill makes him an exceedingly ugly antagonist to man, dog, or almost anything else. It is reported that even Audubon, himself, was threatened by this dangerous Sandhill challenger with the menacing bill.

The Sandhill Crane rests and roosts in shallow waters, usually a sandy hill in an active stream somewhat adjacent to dry open grasslands for foraging. This bird breeds in North America, northeast Siberia, and Florida. It winters in southern North America down as far south as Mexico. These big birds feed on seeds, grain, buds, shoots, leaves, insects, and small animals like mice. As couples, these very social birds begin to shun company during nesting. They arrange a heap of plant matter in tundra, marshes, or grasslands where two or three pale-olive to buff-brown eggs with spots of burnt-umber (over the entire surface) are laid. Both sexes share in brooding, and young learn to fly in about 10 weeks.

The Sandhill Crane has one of nature's most unmistakable calls, a deep triumphant-sounding whoop under, it seems, great effort. The call, which rattles or rolls "*F-a-r-r-r-o K-e-a-r-o-r-r-o*" and is repeated two or three times, is true excitement to fellow flock and man.

My above painting is a scene near the forest road that runs from Ashton, Idaho (US 20) east to Flag Ranch (US 191, 89) halfway between Yellowstone National Park and the Grand Teton National Park. The "sea of flowers" is Mule's Ears; and these wild flowers look like water in the distant meadow. The meadow is called Gibson Meadow. We, who love birds (my fellow-scouter Tom Bird and I), have enjoyed one of the most exciting experiences in all bird watching. For we have spent hours watching the antic, frenzied courtship dance of the Sandhill Crane. What fun do they have springing up high constantly yelping their trilling call! They could have danced all day, and still we would have begged for more! What an experience in birding!

DUCK, MALLARD

The Mallard is most likely the best known duck in the world. This duck looks as though its wings are placed farther back on the body than they really are. They appear so because proportionally the wings are very broad and the tail is quite short. This familiar duck is a modest 20 to 28" in length. The male has a yellow (perhaps a little yellow-green) bill, glossy green head and upper neck, white lower neck band, dark breast, light gray-brown back, black rump and tail, and white under-parts. In winter and when breeding, the tail feathers curl upward. The female has a dirty-orange bill, rather uniform mottled brown plumage, but the head and belly are lighter. Both genders have yellow-orange legs and feet, and when in flight show dark purplish-blue speculum, broadly bordered in front and back with white.

The mallard has been welcomed to "waddle like a duck" around man's home for centuries, and has provided mankind with food (eggs and meat) and warmth (down for clothing and bedding). Duck legs are placed near the rear of its body to help it swim, not walk.

This bird breeds near warm lakes, small rivers, city ponds, and swamps all summer long in the northern hemispheres of America and Eurasia. They winter as far south as Mexico, North Africa, and Southeast Asia.

They feed on remaining grain and grasshoppers in harvest fields. When searching the bottom of shallow waters (using a tails-up bobbing technique), they forage for snails, aquatic insects, fish eggs, and tender aquatic shoots and plants. Their selective diet makes their meat quite delectable. Little wonder we have encouraged their domestication.

Mallards don't mate for life; in fact pairing rights belong to the strongest male. To view the ruthless fight of two males is astonishing: each having a tight pinched grip on the back of the neck of the other, each attempting to drown the other by shoving its rival's head under water. Round and round they go thrashing in the water. I once watched two Mallards fight for hours, each wanting the upper hand (no mandibles) over the other. Eventually one becomes more exhausted than the other; thus a winner is determined. This victory grants him immediate powers over other challengers. A frantic charge (flapping, splashing swimming) towards challengers floating nearby usually keeps them away. After choosing a nest site, 8 to 10 olive-green eggs are laid. The male will stay nearby for a short time but will seek a different sheltered lake for his own safety; the female will brood the clutch for twenty six to thirty days.

Mallards are very vocal for ducks. They call while in air, on land, or in contact with waters. The male utters a loud *rab-rab-rab* kind of quack; the female calls a *quack-wack-wack.* There is much quacking variety from duck to duck, but the female's loud quack really resonates, "This is really a proud mother!"

What an attention grabber - to hear mallards join in a therapeutic but erratic conversation of "quack-quacking". Even the young ducklings seem always talking to each other with a soft *"rab-rab-rab".* We all enjoy conversation, don't we? Mother loved to "talk and walk" with others! Friendships are mentored and strengthened through healthy conversation. Isn't there someone you would like to converse with right now? Why not give twenty minutes of loving conversation to someone right now? They would enjoy your company. However, don't "quack-quack" too long. Remember, twenty minutes is just about right, isn't it?

MEADOWLARK, WESTERN

Meadowlarks are not really larks at all but are more closely related to the orioles or grackles. They are robin-sized birds 8 to 10½″ long. The long and slender look of the meadowlark is caused by its tapered head and slender bill. The Eastern and the Western Meadowlarks look very much alike. If they were perched side by side, only an expert would know they weren't the same species. The Western is browner and the yellow extends slightly farther onto the face than the Eastern. They each are bright yellow below and have dark eye strips and V-shaped black bibs. Their throat and cheeks are yellow, and they have black and white strips on the tops of their heads. Their wings and tail are barred with brown, and their outer tail feathers are white. The legs and feet of both are pinkish.

This bird has a number of little quirks. It flies using two techniques alternately: first flapping, and then sailing. Meadowlarks always walk; they never hop or run, and upon alighting they flutter their tails vigorously, perhaps to lead on someone who is watching. Ornithologists differ on whether Eastern and Western birds mix causing hybrids, but none suggest their different songs are affected. Therefore, it appears to me that this theory fits best. Eastern and

Western Meadowlarks occur together in places; yet no interbreeding happens. They stay divided by affixing to the nature of each species' own song. The meadowlark's insolating mechanism in choosing a partner is its call.

Meadowlarks like open grasslands, pastures, savannas, and marshes, especially when areas are bordered with brush. They feed on the ground, eating seeds, grains, spiders, and other insects. They will use their sharp bill to dig in the soil to acquire their favorites.

They will build a nest on the ground in a concealed spot. They partly dome the nest with grasses, leaving an entrance to the nest on the side. They usually lay 3 to 7 highly spotted white eggs. Both the Eastern and Western species build similar nests and are polygamists, sometimes causing the maintenance of several nests in the same territory.

All of the emotions, exuberance, and movements of a great performer are evident when watching this bird sing. It performs with strong lungs, throat and larynx quivering, head thrown back and projecting its vocal resonance, perhaps to the entire world.

When I was young, I grew up believing the Western, with its flute-like voice, only sang about where I lived, "*Smithfield is a pretty little town*". Later in life when living in the upper mid-west, I learned its song "*See-you,*" then "*See-yeer*" followed by *"Dzzer".* Actually to me, in a whistling-type voice, the Eastern song goes *whe e-hir-weedle-e-whi-chee*. Each species repeats its melody often.

When calling a friend who you have not seen or talked to for a long time and they recognize your voice, doesn't that make you feel good! A lady I had worked with twenty years earlier called me and said, "Bet you don't know who this is!" I answered, "It's Dora!" and she was thrilled. My gal has a delightful laugh that warms the whole room. What a pleasure it is to hear friends laugh and enjoy life together. Do you know someone who is described as having a radio voice? We should use our voice with care, curb temper and tongue, give soft answers, and develop the ability to communicate. The distinction of our voices should be used with the skill of the sweet sounding meadowlark whose voice warmed my heart as a little boy and has continued to lift me (now in maturity) with its song.

EAGLE, BALD

The Bald Eagle is 32 to 40″ long with a wingspan of up to 7¾′, whether sighted in flight or perched. The adult (over 4½ to 5 years old) has a white head and tail. The body, legs, and wings are blackish-brown, while the beak and feet are yellow.

The Bald Eagle is a skilled, powerful hunter. This white-headed eagle has powerful eyesight that can spy the object of his desire from two or three miles away. It can capture waterfowl in flight, rabbits on the run, or snatch fish from beneath the water surface with ease. In fact, it can secure food at will. It gives distinct, but light chattering calls, "*che-che-che*" - "*che-che-che*". Many have honored or used the eagle as a symbol of strength and courage, suggesting others should look up to it as the world has looked up to the eagle. In 1798, the U.S. Congress chose the Bald Eagle over the Wild Turkey as its national emblem. This bird seems to represent dignity, majesty, and pride for cause or country.

Perhaps the largest existing Bald Eagle population is found in Canada's British Columbia and Alaska, nearing 3,000 birds. In addition, the Bald Eagle is found throughout North America and eastern

Canada. These birds migrate south to find waters (not iced) to fish. From December to February, they are a common sight in states about 38 degrees north.

They eat mainly fish, small animals, water birds, and carrion. Because of their skill in finding food, they become providers of provisions for other birds and mammals. We (my wife Patsy and I) one day watched a Bald Eagle devouring a large fish near the Farmington Bay, adjacent to the Great Salt Lake in Utah. Quite a crowd of birds had come to watch and obtain some fish droppings. There was a Red-tailed Hawk, three Black-billed Magpies, and a Northern Shrike all awaiting the dropping of some fish parts by a casual-eating dominant raptor. Each dropping was sought after under nature's pecking order rules. However, quickness at times violated this rule of order.

They make very large nests of sticks on a cliff or in a tree. Bald Eagles pair for life and will return year after year to enlarge the stick nest before laying one to three eggs. The nest will grow each year until, eventually, it becomes too big for the tree to support. The young eaglets soon resemble teenagers. In the first year, the immature eagle will become larger than its parents. Its wingspan may be nearly a foot wider than its progenitors.

As their plumage develops in preparation for flight, they must build muscles and agility to match this feat. So preparation for flight begins by consistently stretching its wings over the nest and performing a jumping-jack kind of wing flapping game. To watch this activity causes fear that they might flap themselves off the nest before they are ready. In their lifetime, they have viewed the expanses of this world from lofty perspectives. They have met perils and prevailed over them. When we sight this white-headed mature adult, a feeling of respect is brought to our lives. Maturity suggests wisdom and earned pride.

MAGPIE, BLACK-BILLED

It's a large, conspicuous, and showy 18 to 20" long bird with a striking long wedge-shaped tail which comprises a little more than half the length of its body. It walks on the roads and ground with a convinced swagger. It has a large bill, sturdy head, short rounded wings, and stout feet. It is black above including head, bill, and eyes. The Black-billed Magpie has a white breast, shoulders, and quite showy wing patches which flash in flight. Tail and wing feathers show glossy iridescent greenish-blue and sometimes purple.

Magpies are a mixture of shyness and boldness. They are quick to get out of the way of an oncoming car but persistently come back for their share of a road kill. Magpies, like jays, crows and ravens, show their intelligence through resourcefulness. Some ornithologists feel they are better thieves than rooks. In typical magpie counting, it is a common sight to see smaller birds attacking magpies as this bird makes its rounds to identify nests unguarded. This bird is quite comfortable near man. Man very rarely recognizes the magpie's helpfulness but always sees his imprudence or rude behavior. Our environment is cleaner because of their persistence in eating every scrap of carrion. Magpies are residence birds, mate for life, and spend summer, winter, spring, and fall in the same territory. They are easily recognized throughout Western North America in slow but graceful lopping short flights. The tail is long. They are unmistakably magpies.

Magpies dwell in sparsely wooded farmlands, ranch lands, mountain lands, and valleys. They fear not and will take up space very near residential or business facilities. This bird is never too far from a stream of water. They nest from southern Alaska throughout prairie states and Rocky Mountain states.

The Black-billed Magpie's taste for flesh and blood is, in fact, helpful. When a badger or skunk becomes a victim of road kill, the magpie, like the vulture, cleans up the stinking mess quickly. Darting in and out, avoiding cars sometimes on very busy streets. It doesn't take long until the stench is gone. As often as we see magpies working on pesky scenes, it becomes hard to believe that their main source of food is insects.

Black crickets, grasshoppers, grubs, and larva are destroyed in huge quantities by magpies. Magpies also eat fruit, berries, and some green leaves. They will patrol nests of other birds and pilfer eggs, as well. Magpies eat about everything!

The nest is a bowl shaped of mud, grasses, and perhaps some bright pieces of twine or ribbon to add color from scraps they have found while strolling their territory.

"What bird was that?" "Oh, it was just a Black-billed Magpie!" How can we give such little respect to one of God's creatures with so much talent? This bird knows when to be patient and when to be aggressive. Most humans are one or the other, but not both. Is it attitude or disposition that makes us the way we are? Do we always wait for just the right moment? Should we not meet the situation with the frame of mind that could steal the scene? Cannot influence be enhanced from time to time with provisional displays of personality? Try it. There is a bit of actor or patient in all of us. Those who study behavior and wish to influence others suggest we must change ourselves to invite modification in those we wish to influence!

AVOCET, AMERICAN

The American Avocet is 16½ to 18½″ in length, spindly legs, and slender upturned bill. The Avocet stands out among North American shore birds as graceful, snooty (up turned bill), showy, and stunning. It is black and white above, and white below. When startled, it cackles its call and reveals its back and wings which are a striking display of stripes or patches. A loud sharp "*week*" or "*kleek*" is often repeated. This cry calls for help and neighboring avocets come and assist in a fury directed at the intruder. It has bright, but dark brown eyes. The head and neck in breeding adults is cinnamon red to chestnut brown (in winter birds, the head and neck appear ghostly white). Its long legs and feet then trailing in a quick response flight are light blue-gray.

The avocet is taller than other shore birds with which they associate. They are often seen standing in the water on one leg [as is shown (left) in my artistic interpretation above] and sometimes sleep that way with the bill and head tucked into plumage or under a wing. Although the American Avocet and the Black-necked Stilt seem to search the same waters for sustenance, they do not compete for food. The needle-like bill of the Black-necked Stilt is used to spear its prey and pick it up with the precision of a pair of tongs. The up-turned bill of the avocet is vigorously swept from side to side constantly through the mud and mosses below the water surface. The upper and lower mandibles are so sensitive that they can open and close on small minnows or invertebrates which they cannot see in the cloudy, mucky waters. Tiny crustaceans, minnows, small invertebrates, and shrimp are sought for. It is one of the birds which will feast on brine shrimp found in places such as the Great Salt Lake in Utah. It will sometimes submerge head, neck, and bill as it finds deep waters full of food, as well. The American Avocet has plumage that is duck-like and can swim when it needs to. It has three webbed toes in front, which are unusual for a wader and one basic hind toe. The nostril in the thin bill is close to the head to avoid water when probing.

The American Avocet is commonly found in the shallow waters of ponds, marshes, and lakes in fresh or alkali water with ponds of mire. It breeds mainly in the Great Basin, Canada, and the Central Valley of California, and in the prairie states.

Its nest is a shallow depression on the ground, sometimes moist or muddy, lined with dry grasses. Four olive colored eggs with brown blotches are laid. Brooding is appropriately spaced from nests of other pairs to provide plenty of range, yet close enough for group support when predators approach.

The female's elegant bill is slightly more upturned than that of the male. Perhaps the pride of a well-dressed lady is keener than that of a handsomely appareled male. Does not this avocet teach us that perhaps it is alright to have your nose in the air just a little bit? Of course, if the nose is not lifted in the air distinctly, people might forget to look up at you at all?

Why would anyone want to stand on one lag in the water? Could it be he is just plain tired of standing in the mud and now wants to get at least one foot dry? Or is he showing off? Remember, he is one with his nose in the air. Well, I don't think so. For when he stands alone like this, he appears serene to me. It is a pose of contentment - elly full, just satisfied with his work for the day. And, to prove my analysis, he will tuck beak, head, or both under his wing and take a nice snooze. Oh, for the bliss of a one lag stand!

CORMORANT, DOUBLE CRESTED

This Cormorant is 30 to 36″ long with a wing span up to 4½′ wide. It is a somewhat elongated water fowl with short, powerful, sturdy legs. For power swimming reasons, the webbed feet extend beyond the base of the tail. In addition to giving swimming strength, they have a peculiar up-right position when perched. The bill is long, straight, and hooked at the tip. They are usually silent, except in nesting colonies where grunting or pig-like sounds are made. It has a comprehensive large head and far-reaching crooked neck. Its wings are pointed and used for swimming, as well as flight. The adults are black with a greenish-black sheen. The immature are brownish-black and have a somewhat lighter breast and neck. Both adults and young have a bright, yellow-orange throat pouch. The Double Crested is distinguished from the other two cormorants by this color. Rarely seen, but present when mating, are the tufts of soft curved feathers extending less than two inches in the back of each eye. These crests are mostly white in western birds; however, they are black and less conspicuous in eastern birds.

The cormorants have a unique and distinct ritual of partnership while incubating the eggs or young. One of the pair is on the nest while the other, at the appropriate time, comes and lovingly circles and nudges the partner. Suddenly, the partner responds and flies off, joining other cormorants in lines or V's as they patrol the surfs for schools of fish. The above painting was created from pictures taken west of Monterey in California. Their constant scan of the Pacific for schools of fish was improving our enjoyment of this stunning seascape. Cormorants seem lacking in oil secretion to protect themselves from the water, for much time is required for preening and drying wings. The striking poses of these birds while accomplishing this self-care are quite eye catching as they perch erect and stretch these appendages of flight against the brilliance of the sea waters.

The Double Crested is quite able to adapt, which is evident because of its inland selections of surroundings. They prefer seacoasts, harbors, lakes, rivers, and large inland bodies of water. To create bird refuge waters, dikes have been used. Double Crested Cormorants can easily be viewed fishing swift waters flowing down between dike water levels. They struggle for position on these streams, like grizzlies would while fishing for salmon.

This bird is a marvelous swimmer, showing spectacular speed on or under the water with the grace and agility of a well-trained athlete. Identifying, targeting, and catching fish between its two mandibles is done with ease. When surfacing with the catch, they will many times thrash the fish to stun it before attempting to toss it up, catch it, and swallow it. Their diet is, obviously, mainly fish.

Cormorants nest in colonies on rocks or in trees. The nest is a mass of sticks, seaweed, moss, earth, feathers, and trash. The male seems to be the supplier of materials and the female is the arranger or builder. Three to five long chalky-blue eggs are laid.

This bird has a number of skills worth imitating. On or near our beaches, they show Muscle Man or Bathing Beauty how to pose, or just plain look good. Their ability to fish, surf, toss, catch, and swallow would put any circus performer to shame. However, the ability to communicate with one's partner through ritual or demonstration is a skill all loving couples could emulate.

BLACKBIRD, YELLOW-HEADED

A late spring snow storm provided a colorful background for this rendition of the Yellow-head.

This bird is a most striking marsh-nesting songbird. The Yellow-headed Blackbird is 8 to 10″ long, which is larger than the Red Wing Blackbird (perhaps robin-sized). The female is noticeably smaller than the male with much duller plumage. The adult male has a bright yellow head and breast with a small black mask which encircles each black eye and extends down to and across the top of the bill. White wing patches are only noticeable in flight and truly enhance sightings. The adult female is dark brown and lacks a wing patch. Her distinct brown eyes, eyebrows, lower cheek and throat are tinged with yellow, and have light streaks of white on her belly. The juvenile are dark brown and edged with tan on their neck, back, and wings.

In courtship display, the male Yellow-head is slightly crowned with red, which seems orange next to the yellow. Wings are outstretched to show the female the bright white wing bars. Bowing to his potential mate, the male Yellow-head puffs up in bluster of feathers and under extreme effort, renders his gurgling, harsh song. He shows the endeavor of a fine musician offering his raspy, rusty door-hinged kind of sound followed by a long, descending "*buzz*". When an alarm is needed, it makes a "*krick*" or *"kack"* sound which is followed again by the raspy notes and "*buzz*".

The Yellow-headed Blackbird is quite gregarious and flocks together with other birds. These flocks in the Southwest become quite immense (thousands or even up to tens of thousands). Large flocks are created for feeding and protection purposes. A study like this may find the flock with no female Yellow-heads. This is because the female usually flies with separate flocks and tends to fly farther south into Mexico, while the groups with male Yellow-heads tend to stay in the southwest United States.

Yellow-headed blackbirds are extremely rare in the east. They are found mostly in western and central North America. They prefer fresh water marshes or lakes with reed beds. They are found from British Colombia to Saskatchewan on the north; to California, southern Arizona, and New Mexico on the south; from Washington, Oregon, and California on the west; and to the Great Lakes and Great Plains in the east. They nest in deep baskets of stems and grass woven onto reeds from 2 to 6′ above water and lay three to five very pale blue eggs with reddish brown blotches. Incubation by the female only takes from 10 to 13 days until they hatch. Nests are carefully distanced from the neighbor for privacy but are close enough for protection through unified alarm. Their diet consists of insects, snails, seeds and grains.

The Yellow-headed Blackbird nests in large dense colonies, and most often the marshy reed beds are filled with their raspy, rhythmic songs. Thousands of these birds, like a large curious clucking, gurgling, buzzing community, join together in this strange chorus of unity. It is a unique hymn of defense, a rhythmic melody of cooperation, and a romantic ballad of fellowship. To hear this illustrious choir begin to sing at dusk, ease your car as close to the marsh as possible, prepare yourself for mosquitoes, roll the windows down and listen. All of the gurgling, purring, and rasping sounds might suggest the symphony orchestra is tuning up and soon will play a magnificent piece of music. But, the tune up never ends; it just goes on and on in a somewhat melodious never ending tune up. I cannot attest that the pre-musical adjustment never ends; for soon enough is enough, and we go home.

DOVE, ROCK

This dove is 12 to 13½″ long. The domestic pigeon is spoken of in many ancient cultures but was introduced to North America from Europe as a pet and as a source of food and feathers. It has a round head, pointed wings, fantail, and is always in flocks. This dove can be kept nearby through consistent feedings. In the wild, they feed on grains and other seeds. In cities and parks, they subside on whatever man wishes to feed them or is available as refuge scraps, crackers, bread crusts, or other baked goods. There is in man a satisfaction in feeding pigeons.

The above artist painting was painted from a photo I took at La Jolla Beach and Park just north of San Diego, California. These pigeons gather morning and evening to be fed daily by tourists congregating to view this beautiful ocean scene; but finding these attractive friendly doves, they feed them. There are numerous color variants with bright iridescence on the neck and shoulders (blackish, reddish, and greenish variant color). Rock Doves are typically gray, whitish, or piebald with two black wing bars. They often show white rump with a black band at the end of the tail. Rock Dove plumage at close range is quite colorful, but when viewed from a distance is not particularly striking.

Rock Doves are gregarious birds, moving about in flocks during the day and roosting together at night. Disturbances between birds are quite seldom. When disagreements do occur (usually when breeding) the bluster is really for show only. Rock Doves begin each day with a chorus of cooing calls, followed by flock ventures for feed and drink. In the heat of the day, resting time is taken. Once again in the afternoon they seek food and water. These routine habits and their friendly nature have always made this species close to man. Their amiable attitude and realization that man means food has led to their domestication.

Rock Doves live in cities, parks, gardens, suburbs, farm areas, rocky canyons, and sea cliffs. They are found throughout southern Canada and the states. In the wild, they feed on grains and other seeds. In cities and parks, they subside on whatever man wishes to feed them or what is available as refuge, such as seeds or grains, bread crusts, or other baked goods. There is in man a satisfaction in feeding pigeons. This bird does not migrate.

A flimsy shallow saucer is made from a nest of sticks on a window ledge or cliff. Two white eggs are laid, and each sex takes time for brooding. When chicks are hatched, both parents feed them "crop milk" from an immersed bill, which is also used for drinking.

Their song is a soft, deep cooing *coo-cuk-coo-cuk-coo.* The cooing of the adult male and female is truly a melancholy, loving sound. Perhaps the Rock Dove is the ultimate example of love, courtship, and affection. "Lovey dovey" stuff is perhaps a phrase we have heard many times, as a put-down of sweet, real emotional attraction or love. However, it is truly correct about this bird's kind disposition. The male's deep cooing song and his display of drum major type strut of pride is shown to the female. Each pair mate for life. As well, the Rock Doves are very sociable and loving as a community. As the pair's coos sound together, it seems like real conversation expressing fondness and love for each other. Long periods of just being together, not speaking, have been observed. It is followed by a stroking or petting of the feathers, male to female, across the head and down the back of the neck. Yes, affection and love are indeed exemplified by a pair of Rock Doves.

HUMMINGBIRD, BLACK-CHINNED

The Black-chinned is a small hummer 3¼ to 3¾″ long. The bill is moderately long and slightly de-curved. The adult male has a black chin, bordered just below by iridescent purple. The gorget is made distinctive by a near-white collar. Crown, back, tail, and wings are green. There is a white spot behind eyes, and the upper sides of the neck and breast are somewhat iridescent. The under parts are whitish, and the tail is notched. The female is not distinctive. It has green above and white beneath, and the flanks are tinged with buff.

Its song is a soft "*chew*", followed by very high buzzing notes.

They usually seek out broadleaf trees to nest in. The nest is an intricate cup of plant down, lichen, and feathers bound together with cobwebs. After a swift mating, the female bears full responsibility for nest building. She uses a quivering motion in order to mold the cup properly to her cozy liking. Two white eggs yield chicks, which are fed digested nectar bill to bill.

These delightful hummers are found from British Colombia south to California, and from Montana to upper Arizona, and in northern New Mexico and Texas in the south.

HUMMINGBIRD, ANNA'S

This husky hummer is just medium-sized 3½ to 4″ long and has a needle-like beak. The adult male is mainly iridescent green above with a rose-red crown and gorget. It is the only North American hummer that has a rose-red crown. The throat patch is elongated on the sides of the neck. Its tail is black with gray-green below. Females have rose-red iridescent flakes on the throat. The incidence of light reflecting from these spots can cause them to disappear, change colors, or be quite bright. Dingy gray is underneath and the tail corners are white.

They sing from a perch and make a sharp "*chip*" followed by a series of "*chee-chee-chee*". The male gives a high pitched squeak-like "*te-uit, te-uit*" while performing display dives for his chosen mate.

Many feel this is a California bird which breeds from Baja, California north to the Oregon border. After breeding season they migrate in search of nectar. They fly north to British Columbia and Alaska, and south to east Texas, and have recently been found in northern Arizona and the Rocky Mountains.

HUMMINGBIRD, RUBY-THROATED

This is a small bird, 3¼ to 3½″ long. The bill is somewhat long and straight. The tail is deeply forked. The adult male is iridescent green above, including the crown with a brilliant red-ruby gorget, and flanks tinged with iridescent olive green and a dark forked tail. The incidence of reflection from the iridescent crown and gorget as well seems to continuously change in colors. The female lacks brilliance; it is somewhat green iridescent above and white below.

A thin high "*chick*" call and some

other squeaking noises make up this bird's song.

The nest is less than 2" across on a branch decorated with lichens and held together with spider webs. The dully-plumaged female bears the burden of building the nest, brooding, and feeding the young.

The Ruby-Throated Hummingbird breeds in North America, from Alberta to Manitoba, and south through the Great Plains, Texas and Florida. It flies south in winter to southern Florida, Mexico and Central America. This 500 mile journey across the gulf is made without food or rest twice a year.

The best way to observe these colored little birds is at a feeder. They love sugar water feeders, which should be maintained and kept clean. The mixture of sugar needs to be weakened before cold weather comes, or hummers can be bated into freezing weather by mistake. However, in good weather a feeder provides a delightful way to observe these wonderful little birds. They will feed on as many as a thousand blossoms a day. Hummingbirds have long grooved tongues that are usually forked at the tip, allowing them to lap up nectar from flowers. They also feed on insects that the flower nectar attracts. The insects included in their diet are ants, aphids, gnats, and spiders.

There are over 300 species of hummingbirds and all claim similar distinctions. Their display techniques include brilliant iridescent colors, high demand for energy, and unique capabilities to fly in every direction. Humming birds use energy very quickly and must feed about every four to ten minutes throughout the day. These small birds have relatively large hearts which beat five-hundred times per minute, constantly pumping fuel from the digestive system to the muscles. When feeding areas are found, somehow many hummers seem to be alerted. Thus, awareness requires one of these fierce little hummers to become a bully. Energy seems then to be wasted as a vicious game of dart-like war is begun to claim all the nectar for oneself. Much patience is required by the fellow hummers to get a sip or two of the sweet nectar.

Indeed, the hummers are captivating, amusing, and amazing in flight. The pointed, fixed wings (no real movement at wrist or elbow) attached to the shoulder by a swivel joint can rotate 180 degrees, and each feather can revolve on a plane, like Venetian blinds. The tips of the wings draw rapid figure eights in the air as they perform their feats of skill. Using all of these features, the hummer flies backwards, frontwards, straight up, and can drop down by turning the feathers and allowing the air to pass through. Wings sculling rapidly back and forth provide all of the dexterity of helicopters and more. The hummingbirds have supreme mastery of movement. Their wee feet and legs are almost never used, owing to their great dexterity in flight, the only exception being rest at night.

They are quite inquisitive in nature and can surprise you by buzzing close to humans, quite out of curiosity. The male hummingbird provides remarkable diving acrobatics as part of the bright, colorful display to attract a suitable female. From as high as 100 feet, The Ruby-throated hummingbird executes pendulum performances, distinctive display flights of half circles and dives, always before a perched, seemingly unimpressed, female. The Anna's give long, steep dives with an explosive popping sound given at the bottom of the decent. The Black-chinned can be observed performing shallow pendulum or swinging display flights. All of the demonstrating shows talent and skill. These shows are for enjoyment and are always done to impress or woo a mate.

We, surely in this life, should learn and remember that there are many ways to accomplish things. We should not get caught with our minds in a box. Habit or tradition usually is quite confining. We should break free and try things a new way once in a while. It could be that we should visualize the dexterity of the hummingbird when faced with a challenge and see ourselves doing things in a different way. Perhaps, a hover to surmise, a quick uplift or down drop to obtain new perspective; or for all we know, a problem could be reviewed backwards to find the errors. Yes, hummers can teach us to approach life differently.

Dexterity, handiness, nimbleness, and proficiency all must be admired in the flight of the hummingbird. Speed, dispatch, hustle, and quickness are surely attributes and demands of this busy world. Yet, we must learn when and how important it is to just plain shut off work and worry, and just plain rest. If the hummingbird did not become torpid (the heart beat slowed, body temperature lowered) each night, death would come. We should learn to find deep rehabilitating rest which, of course, blesses our lives with the ability to work and produce. We should work and rest like a hummingbird.

PELICAN, BROWN

A spectacular and most interesting great bird, the Brown Pelican is 44 to 54″ long, has a wingspan 6 to 7', a huge bill, and extensive throat pouch. It flies with its head back on its shoulders with its long bill partially resting on its S-shaped neck. It has a distinctive flight of a few strong flaps followed by a glide. The adult's white head and neck are often washed with yellow, the body is grayish-brown, the bill is a number of colors (bone-pink, brown, yellow and blackish), and the feet are silver-blackish. The Eastern's pouch is brown. The Western is reddish-brown with a light eye. Both sexes are alike. Breeding birds' hind neck and nap are cinnamon brown. A yellow patch appears at the base of the forehead. The Immature has a dark head and is pale below. The head, neck, wings, and back are brown. The bases of the neck and under parts are white. Young pelican produce groaning sounds; adults are generally silent.

Like many huge birds, much effort is given to obtain flight. The Brown Pelican always faces the wind to take off. They flap their huge wings and paddle with their webbed feet. When taking off from a pearl perch, the angle of flight seems downward, then slowly upward. When flight is obtained, they are found to be powerful in the air and are great at fishing. They feed exclusively on fish and shrimp. The sleeping Brown Pelican is extremely sensitive to tide waters, relaxing and sleeping during low tide and hunting for fish during high tide.

Their highly specialized bill with remarkable pouch below allows them to capture fish seemingly with ease. Having abandoned the strategy of fishing on the surface, instead they scan the surf zones just off shore working together in long, single-file lines low over the water. When a large school of herring is spotted, the diving begins from heights of 30 to 50′. As the Brown Pelicans splash (looking clumsy) into the water, the lower pouch mandible stretches out like a net to catch the fish.

Oceans' inshore waters are where these birds can be found. They like to rest on cliffs, piers, or pylons where fishing boats are moored. The above painting was created from a photo taken at La Jolla, California, looking down on a Pacific coast cliff. They are coastal birds from the Pacific coast, California to Washington (rarely in British Columbia), Atlantic Coast, New England, and Nova Scotia, south to Florida and the Gulf Coast.

When trees are available, such as mangrove trees, they will build a large nest of sticks and reeds. Two to three white eggs with quite different chalk-white shells rest in the nest. The white feathered chicks are, indeed, quite helpless.

The Brown Pelican, whether young or old, suggests to me lessons on how to manage retirement. They certainly know how to get comfortable, how to relax in the sun, how to yawn and stretch, and when to fly off and go fishing. This Brown Pelican with his white head seems skilled at being friendly or tolerant with humans and obtaining some small favors or handouts from tourists and fishermen. Men in years of privacy and rest seek favors, as well: a senior citizen's discount or a handicapped parking place. When the Brown Pelican has been successful fishing, there is always a gull ready to snatch his take as the fish is carefully flipped or turned in preparation to put it away. So is retirement full of its dangers; for there are many who wish to prey (double billing of the aged, a common practice) on those weakened by age. We can learn much about good retirement from the Brown Pelican, a beautiful big bird who loves a daytime nap.

JAY, BLUE

The Blue Jay is a large bluebird, 11 to 12½″ long. It's a handsome, noisy, and aggressive bird with a conspicuous crest. The wings and tail are about equal in length and rounded. This bird is the only jay in the greater part of eastern North America. It has a grayish or violet-blue crown, crest, back of neck, shoulders, rump, upper tail coverts, and middle wing coverts with pale bluish-gray sides of the head, chin and throat. It has a black necklace and base of crest. It is bright blue barred with black greater wing coverts, secondaries, and tail feathers. Wing feathers are blotched with white and tail feathers are broadly tipped with white. The breast, side, and flank are gray and the abdomen and under tail feathers are white.

In addition to the beautiful coat of feathers, this jay sings a warbler-like song of his own and mimics other sweet and melodious bird songs. It's known for its harsh voice. Here are a few descriptions of the sounds: *"eeef-eeef* or *thief-thief"*. Another reports *"jeeah-jeeah* or *jay-jay"*. Why he chooses to mimic is not known, but most would choose to believe it is for deception; for this bird is a rascal full of mischief, deception, and thieving. Blue Jays flock in small groups and go marauding with their own sharp crow-like *Jay* call, as well as their imitation calls of raptors, such as the Red-shouldered Hawk. This tactic is used to scare other birds away and plunder the area. They will identify nests in the surrounding forests and make rounds to check the homes of other birds in hopes of finding chicks or eggs unguarded. They seem to find delight in pillaging nests and devouring their contents. This bird sounds a very noisy alarm of "*Jay, Jay*, and *Jay"* when man enters the woods, but in spite of a bright blue coat with black and white adornment, this jay is quite elusive and hides quickly.

Its nest is a bulky cup of moss, twigs, pine needles, and leaves in the crotch of a tree limb 5 to 50′ above the ground. A Blue Jay will lay three to six olive-greenish (spotted with brown or gray) eggs. They are omnivorous, feeding on acorns (their favorite), chestnuts, added grains, other birds' eggs and chicks, rodents and even carrion.

Incubation takes 16 to 18 days and the young leave the nest 17 to 20 days after hatching. The male dutifully feeds the female seeds and nuts during brooding. It lives in mixed deciduous and coniferous forests, parks, and gardens in eastern North America from Alberta to Newfoundland in Canada and from eastern North and South Dakota, and Nebraska to the Atlantic Ocean in the United States; but, it does not breed in the Florida peninsula.

As people come to our home, it is not long until the discussion includes our love for birds. It matters not who -neighbors, people from the church, family, and especially our grandchildren; we will talk about our feathered friends. Our home's walls are covered with paintings of birds; and because of this fact, all who come seem to want to choose which bird painting they like best. Some of the favorites are the Snowy Egret, Goldfinch, Western Grebe, Mountain Bluebird, or one of the Hummingbirds. However, so very many choose the Blue Jay. We explain what a rascal this bird is, but the bright white, black, and blue color seems to grab them. To reward an old cliché: "We should beware of raptors who look like Blue Jays" for beneath that color is a bandit. Beauty should enhance truth, not hide treachery. This marauder is truly costumed to defer its intent to thieve; but, what a good-looking bird he is, isn't he?

FLICKER, RED-SHAFTED

Ornithologists are not consistent about whether or not there are three species of flickers. No matter; this bird loves to eat on the ground and from bird feeders.

However, because our viewing of this bird often sees him pecking, eating, and forming whole cavities in trees, we will imply that the flicker is a woodpecker and that there are three different populations of flickers. We find in the east, the Yellow-shafted; in the southwest, the Gilded; and in the west, the Red-Shafted. `The Red-shafted is a large woodpecker, 11 to 14" long, and the most popular of woodpeckers. It is a distinctively bright and attractive bird.

Flight is undulating (several quick beats followed by a pause or glide). The primary, secondary, greater primary coverts and outer tail feathers are broadly edged with their distinct color: red, yellow, or gold matching the species. This is a climbing bird which has a stiff, spiny tail, used as a prop for climbing trees or poles. Flickers are important in their woodland environment, providing nesting cavities for many other hole-nesting birds. As other woodpeckers, they sound out trees and poles before working to build a home cavity. Of course, weakened trees are chosen. They build their own nest-cavity in trees. They lay and brood five to ten white eggs.

The Red-shafted has a brown-orange head with a gray-blue cap, dark bill, black crested bib, red mustache, brown barred back, and a white rump. The tail is black tipped, spotted black on the gray under parts. The under-wing and tail are salmon-red. The Yellow-shafted has a gray crown with a red nape patch and gray bill. The under parts are spotted black on tan, yellow-brown under wings and tail, and the wing lining is yellow. The Gilded Flicker has a cinnamon colored head, dark bill, red mustache, light gray-brown barred back, black crested bib, and white rump. The tail is black tipped, spotted black on cream under parts, creamy yellow under wings and tail with a yellow wing lining.

Their disposition is somewhat friendly and open with other birds.

They are widespread in all habitats where there are parks, suburban areas, forests, and deserts where large species of trees or cactus plants grow. They can be found from Alaska, across Canada to Quebec, south through North America to the middle of America and the southern tip of Florida.

They seem to feed on the ground a greater part of the time, preferring to eat ants. They consume more ants that any other North American bird. They do take other insects and in winter, some berries, fruits, and seeds. We have seen them dining on frozen apples in the fall and seeds from our bird feeder in the winter.

Their song is a delightful, happy drumming long loud territorial call: *"flicka-flicka-flicka"* or *"whicki-whicki-whicki"* lasting, it seems, longer than a bird that size could manage to sing. Many tones echoed by another flicker's territorial call somewhere in the not too near distant woods. To hear flickers thrillingly pronounce ownership of territory to each other is a reckoning thrill. Conceivably, we should each attempt to express the delight and happiness of life, like the flicker. Have you ever wanted to find a high place and shout about what a great life this is? Yes, perhaps we should just let loose and shout for joy! I always feel lifted and happy when I hear the flicker's, *"whicki-whicki-whicki-whicki-whicki-whicki-whicki",* from high in the tree tops; don't you?

QUAIL, GAMBEL`S

This is a ground dwelling chicken-like bird. This delicately small and pretty bird is about 10 to 11 ½″ long. Wings are short but effective. Legs are stout and moderately long. The head is erect and plumed and the tail short. The male is gray above, the sides are chestnut with white strips, and the crown is rust-brown. White fringe is on the cap, and it has a dark masked face that is also white fringed with a curved teardrop topknot or plume. It has a light yellow belly with a large black patch, which is a distinction. The female is duller than the male. It has a light chin, no belly patch, and the plume is slightly shorter. Otherwise they are quite alike.

Three different songs can be heard. In the spring, quail are found not in coveys (many family groups) but as singles desiring to pair. The male will find a prominent place to perch and crow, "*Kway-o,"* followed by a "*Chi-quer-go,"* until he finds a mate. With hen and chicks, or when the covey is flushed or startled into flight, they will give a loud "*Chi-kar-ke-cut-cut".* However, when hiking along a trail in line or when roosting, they sing a cooing kind of song, a low "*Cha-*

chea-cha-chea".

Quail are capable of thriving in the harsh territories of the southwest but must drink each night and morning.

After chicks hatch usually a dozen or more, the parents lead their young on daily learning tours. They learn to find food; they teach them when and how to hide; they practice them in the skill of running in lines through the flora and shrubbery of their domain, and they discipline these fledglings to seldom use flight. If predators are more scarce (cats, dogs, etc.), they can be lured to gardens or backyards with seeds. They eat seeds, fruits, seedpods, beans, grasses, ants, grasshoppers, and other insects. In winter and early spring, among their favorites are the seed pods of wild rose bushes and mesquite beans.

They live in desert thickets, streamside woodlands, cactus, mesquite and chaparral areas. The Gambel's is found only in the interior of the southwest while its cousin, the California Quail, favor the coastal regions of the pacific southwest. They are found in the deserts of southern California, southern Nevada, Utah, Arizona, New Mexico, western Colorado, and west Texas.

Nests are made in a hollow in the ground beneath a bush, mesquite, cactus, or a tuft of grass thinly skirted with feathers or leaves. Most ground birds are considered game foul, as well; but man is only one source of user. For it is common for up to 50% of eggs or chicks to be taken by predators. However, clutches are large, 9 to 18 buff-white and irregularly speckled with dark brown. Quail can walk as soon as they are hatched and abandon the egg shell clutch site quickly. A clutch of eggs can take weeks to lay; and if marauded, the pair will usually try again in the same season.

Near homes at eventide, it is not hard to find a covey of Gambel's Quail marching in lines like practicing drum majors. Then they carefully surround a source of water where they gather to drink their fill. The chatter is continuous; for this is a social part of each day. One by one they then lift themselves by wing into a thickly covered shrub like a huge wild rose bush; thus, finding protection from storm or predator. Their gentle "Cha-chea-cha-chea" voice conversation continues until sleep comes. When life is good, should we not do the same? Refresh ourselves, review the day with those we love, and find reinvigorating rest. Wow! What a way to conclude a day.

GULL, HERRING

This is the most common large gull found in North America and is, indeed, distinguished by its size, 23 to 26″ long. They nest using a simple depression on the ground, arranging a few grasses and weeds around. They nest in colonies usually on coastal islands, laying two to three pale green eggs with brown blotches.

This gull is well prepared for all weathers, vicious storms, and lakes or seas. The head, neck, tail, and upper parts are pure white with a gray-blue mantle and wings. It has black wing tips (some primaries and secondary have white tips). The bill is yellow with a red spot on the lower mandible near the tip. The legs are pinkish; the pale yellow eye is almost whitish with a yellow eyelid ring. In winter, the adults have extensive dusk molting on their heads and neck. In the juvenile's first winter, it is brownish with a blackish bill; the tail and wing tips are brownish-black. In the second winter some blue-gray appears on the back, but mostly has light brown molting. The tail and wing tips are dark brown. Some black still appears on the bill tip.

They dwell on sea coasts, bays, rivers, lakes, and ponds. They can be heard screaming a long, "*kee yow-kyow-kyow-kyow*" sound as they search inland, over seas, or along coastal shores for food. They live in North America (including Canada and Alaska), the Great Lakes, and Long Island. These birds winter along Atlantic and Pacific coasts, as well as Mexico and the Gulf regions.

They eat fish, small marine animals, eggs, and young of other sea birds, as well as refuse. They delight in feeding on live fish, such as herring fish, which are available in schools near the water's surface. This common practice reveals the source of their naming. They are also skilled at breaking the defensive covering of shellfish, such as mussels and clams. They take the shellfish to proper heights and bomb them on rocks to expose the meal.

On the wing, they show ease and distinction as they use their long, narrow sail-serviced supporting appendages to wander over water with no difficulty. They use their alula feathers like other birds and aircraft to ease their decent as they salvage morsels of food from the water's surface. In addition to these amazing forward and soaring flights, they hang right about in the sky with ease and can move forward or backward while hovering. After scavenging far inland, these amazing gulls show formation flight when heading home to roost. Their populations are on the increase, perhaps because of the constant increase of refuse made available by man, and their unique ability to survive. Their life span exceeds that of warblers, sparrows, robins, and even Canadian Geese. They usually live nearly five years, but exceptions have been recorded showing some have lived as long as 30 years.

Never ending pursuit describes this gull. The Herring Gull in flocks pursues schools of herring and dines at will. The schools of herring in some way detect their losses and defend the fish school by swimming deeper in the seas. The herring don't stay deep in the water long, though. For some reason, they just naturally want to swim near the surface of the sea water so the flocks of Herring Gulls make wide circles, thus commencing their pursuits of the school of herring fish once more; then begin all over again to feast. Perhaps, we as well should circle about and start anew at times, as these wise birds do. Starting anew at responsibilities we have yet to complete is good. New vigor, a new approach is when persistence wins, isn't it?

EGRET, GREAT

This Common Egret is 36 to 41″ long with a wing span of 4¼′. It is a wader that shows great muscle strength in breast and neck. It has a stout long bill and stands as tall as 30″in the water. It flies with head and neck folded back towards its shoulders.

This great bird is the symbol of the Audubon Society and displays great white plumes during mating seasons. With white above and below, both sexes show this lacy white veil or gown display in season. It has a mighty bright yellow bill which yields a song that is a low guttural growl or "*crooke*". The legs and feet are dark black. The White Heron, found in Florida, is larger and has yellow legs and feet. The Snowy Egret is smaller and has dark legs and bright yellow feet.

They eat small fish, crayfish, frogs, snakes, insects, and small near-water mammals.

The Great Egret makes its home in lakes, lagoons, marshes, along rivers and coast lines in brackish fresh or salt waters. This large heron is found on all continents, in Europe (near the Danube) and in southern

Russia. It is most common and widespread in North America. The largest of the four egrets is found in Australia. In North America, it ranges in the west from south Oregon and Idaho, southwestern states and down to Colombia. It also ranges from east Minnesota and south Ontario to Texas and Louisiana and also along the Atlantic from Long Island to Georgia and Florida.

It is usually seen almost motionless or stealthily pursuing prey while elegantly showing the beauty of a slow motion ballet. The sheer loveliness of the plumes or egrets of this bird seemed sure to cause its extinction in the late 1800's and early 1900's. Ladies' stylish hats appeared to demand slaughter of the Great Egret right at the time each year that eggs or young needed attention from both sexes of the pair. In 1902, it was calculated that over 200,000 birds were killed, not counting eggs or young that were destroyed. This was done to satisfy London hat designers. At about this same time, Egret fancy feathers exceeded the price of gold per ounce, and the Audubon Society began a campaign to stop the slaughter of this elegant fowl.

Greeting or courtship ceremonies of this creature include calls, neck stretching, skyward bill pointing, and mandible clicking. The sexes share brooding and clutch tending and a unique ritual of passing the baton (a stick or twig) has often been observed as a part of changing the guard of the nest. The twig is given to the mate, then the giver steps out of the nest and the receiver steps over the abode and arranges the stick to enhance the domain. The platforms of stick nests are in tree colonies 20 to 40′ above ground or among reed bogs where they lay three to five pale blue eggs.

The cost of elegance should, perhaps, be carefully evaluated. Yes, femininity deserves adornment, but when embellishment is abused to gratify pride, it can be too much. As the picturesque splendor of the great plumed egret is found in nature, so should we not also find grace and loveliness in the purity, cleanliness, and loveliness of an attractive woman? Yes, some trimming is good, but her true beauty should shine forth first!

GULL, CALIFORNIA

The painting above depicts Antelope Island and the Great Salt Lake in Utah with the State Bird in the foreground.

The California Sea Gull is a medium sized gull, 20 to 33″ long, just slightly larger than the Ring-bill and yet smaller than the Herring Gull.

The name California Sea Gull seems really to be a misnomer, for this species spends most of its time far inland. This bird seems delighted to take on a dirty or horrendous job in defense of agriculture. In flight, they show ease and dexterity along with remarkable quickness and are capable of flying very long distances. Amazingly, and almost mysteriously, they come far inland to totally vanquish field mice and black crickets, as well as grasshoppers in the Great Basin states. Early settlers in the vast sink extol the astonishing arrivals of California Gulls and claim they have been beautifully sent to provide service to our struggling pioneers. Following one of these savage victories over the unwanted, this beautiful bird will return and remove the muck and grime of battle in a frolicking, delightful, community bird bath. As they bathe, they vocalize a wavering repetitive, somewhat soft, "*kee-kaw-yah,*" song which is heard in flight, as well. To witness one of these festive bird wash events makes one ask, "Are these the same warriors just seen in battle?"

The California Gull is found in north prairie states, east to the Dakotas, south to Utah and Wyoming, and west to California and Oregon. They are sighted on coasts, inlands lakes, farmlands, and urban areas.

The adult has a white head, neck, breast, and underparts, and medium-gray mantle. In winter, the neck and head show some dusk markings. Its bill is yellow with a red on black spot close to the tip on lower mandible. Black spots fade during breeding. The legs of breeding adults are quite bright yellow-green. Its dark eyes have a red ring. The wing tips are blacker than other similar species. The juvenile's bill is black at the tip with a gray-pink base. The mantle is very dark overall with brown and black flakes. The underparts are paler and the legs are pink, making distinctions from other gull species very difficult. On the second year bird, the mantle is now extremely gray with light gray-brown below. Primaries are dark and secondaries have a dusk band. The legs are mostly gray, the tail is dusk-brown or blackish brown, and the rump is barred brown and white. The bill is tipped black on a gray pink base.

This gull eats insects, worms (loves angle-worms), rodents, fish (mostly fish scraps following boats), and other leftovers or refuse.

This gull breeds in colonies primarily on islands and the shores of inland lakes, and streams. Colonies are often shared with Ring-billed Gulls. The nest is a cup of stems and weeds on the ground near the water. Two to three heavily blotched and buff-olive colored eggs are laid. Both sexes incubate the eggs, which hatch in 23 to 27 days.

There are many unrelenting services ascribed to the California Gull: huge black crickets devoured then regurgitated in Utah, swarms of locust eaten in the West Great Basin, and herds of small rodents killed in Nevada. Our pioneers attributed it to divine intervention. "You mean that bird that likes trash?" Well, yes. But, you must also say, "Yes, that bird sent by God to rid man from insect storms or waves of rodents, either of which could lie in devastation, lands and lives which now flourish and blossom." As we study gulls, can't we still believe in miracles?

BUNTING, LAZULI

The Lazuli is 4½ to 6″ long. It is a small bird which has a short conical bill. It is a brightly colored finch and is the western counterpart to the Indigo Bunting. It's quite comparable to other buntings in behavior; for it twitches and spreads its tail when perched.

The Lazuli makes its home in hillsides or woodland clearings near water from British Columbia to the west of the Dakotas in the north, and southern California east through Colorado. In the mid-west where Lazuli and Indigo Bunting sometimes breed together, it is not uncommon to find hybrids.

The male has a bright turquoise blue head, throat, and shoulders with orange-brown breast and flanks, sand-tan belly, and bold white wing bars on bright blue wings. Perhaps because of its exquisite blue-headed brightness, it could be referred to as a Blue-headed Finch. The female's upper parts are grayish-brown and are pale below with faint breast streaking and bold white wing bars. A blue tinge can be seen on primaries, rump and tail. Juveniles are patchy and have brown wing bars.

This bird renders long, sweet notes in its songs, sometimes lasting from four to eight minutes; and always near the end, the notes are more rapid and jumbled as they rise and fall. Then they end with a hard "*chip*" and "*buzz-zzd*" sound.

Playful singing and feeding fests seem to be their distinction. In a recent sighting, six Lazuli Buntings were at play. It seemed they would stop on the long grass branch which grew on the edge of Farmington Creek and bounce on the flexible stems while feeding on the seeds at the grass's crown. Some would bounce with excitement until the long grasses would bow to the ground or dip their crest in the stream. Then off they would fly to a shrub of small trees high on the banks of the stream, which is quite steep. The scene these bunting had chosen for rest and play was indeed impressive. The creek was strong, budding, and always moving forward. The grasses were full grown and seeded at the Apex, Poplar, Box Elder, and Scrub or Gamble Oak arching over the stream and framing its beauty.

Their summer diet is about half and half of insects and seeds for adult birds. Buntings, like other seed-eating birds, start their newly hatched young out on insects, and then gradually make the shift to cracked seeds to enrich their diet. The bunting's stout conical bill is designed for seed cracking and mashing.

The female seems most interested in finding a partner and breeding. In fact, if for some reason she were to lose her mate, she would quickly find another and begin the breeding process anew. She also bears the responsibility to brood and raise the chicks. Breeding takes place from April to June. Three to four pale blue eggs are laid in a neatly woven cup shaped nest of grasses and are placed in a shrub or tree from 3 to 10' above ground.

Lazuli Buntings provide quite a sighting! For I know of no time that ever stopped my progress along a brisk mountain creek with more enjoyment! What a sophistication of life! What a recreational display of their agility and nimbleness in using nature's natural playground for fun.

ROADRUNNER, GREATER

This special unmistakable (no similar species) spunky chaparral cock is considered a part of the Cuckoo family. This bird, made famous as a cartoon character, is quite a large ground dwelling bird about 20 to 24″ long with a heavy bill, bushy crest on back of head, and short rounded wings. It has short, but strong, rapped running legs and a large, long tail. It makes a low somewhat dove-like (6 to 8 notes) cooing sound mostly when perched, "*coo-cooo-ooch-oooh,*" each note longer and lower in tone.

This is a well camouflaged bird. Its color and markings blend right into its chosen habitat. Roadrunners, like most desert birds, reduce their activity in the heat of the day. These two factors, plus its high speed (runs up to 15 mph) and excellent mobility make sightings almost always a brief, but exciting experience. They use their tail and wings to aid their sharp turns, quick stops, and darting maneuvers. When discovered or alarmed while perched, they will cock their tail and fan their crest.

The Roadrunner is brown with a green sheen streaked with blue-black and white above and buff below with brown streaked on its breast. There are red rings around the eyes, distinct blue and red naked skin patches behind the eye of both male and female, and long tail feathers. The primaries show white crests.

They are found chiefly in arid or semi-arid ranges, in southwest deserts, open woodlands, chaparral forests, or mesquite groves and farmlands. Because of the cold nights, they seek the warmth of day by finding high spots to perch and sunbathe. One day in Palm Springs, California, we enjoyed a pleasant sighting. From the hotel breakfast patio Roadrunners had found warm sun on the guest house roof. They were quite content up there until the day got quite heated.

They breed in Lower California, Baja California, upper Mexico, bottom of Utah, and in Nevada, New México, Texas and Oklahoma.

These ground-dwelling overgrown Cuckoos run down and eat insects, scorpions, lizards, young ground birds, rodents, larva, fruits and snakes (even the poisonous rattlers). There is a minute indication among ornithologists that the inclusion of small ground birds in their diet is more fable than real; perhaps that is so.

They lay about two to six chalky white eggs in a nest of sticks on a shrub or cactus. The young roadrunners hatch at quite long intervals apart. It is not uncommon to find one fresh egg, one ready to hatch, one or two featherless black-skinned young birds, and two more grown birds about ready to start running around the nest. If food is scarce the large juveniles survive and the small ones starve to death. A pair of Roadrunners lives on their chosen land year round; they don't migrate.

There are many myths and tales about the cunning, intelligence, and unique character of this unmistakable bird. They have the ability to place a fence of cactus around a sleeping rattlesnake, then alarm it to make it strike, thus gaining the use of the cactus spines to aid in the kill of a prospective meal, as well as the unique over dramatization of the beep-beep cartoon character. The Roadrunner has indeed become a famous hero. What an example to each of us of zeal, excitement, and clever craftiness. As we plan, as we envision the future, let us put in our mind's eye the enthusiastic, fervent, and artful spirit of this sporty desert dweller. A run at life, like the Roadrunner, we might try. What do you say?

KINGFISHER, BELTED

The Belted Kingfisher is a medium-sized bird, 11 to 13 ½″ long. It has a large head with a sharp, stout bill. The crest is shaggy or appears bushy. This is the only kingfisher in North America with a belt.

It is the most well-known or profuse kingfisher in North America. It's found near clear waters where adjacent trees, shrubs, or poles are used to perch and spy the waters. It makes the sound of a loud chattering dry rattle or "*caar*" notes given from perch or flight.

In the north, they find homes from Alaska to Newfoundland, and in the south from the Gulf Coast to California. Their winter months are mostly spent in the southern United States.

The nest is in the ground at the end of a tunnel 3 to 7 feet long. They burrow in river banks using their husky beaks, legs, and feet to excavate. The pair shares incubation of five to eight white unmarked eggs.

The male has a blue-gray crest, back, wings, and tail. The collar and belly are white. He has a blue-gray belt across his breast. The female has two belts. The upper chest is blue-gray, and the belly is rust colored, and extends down the flank. The tail is barred with slate and white.

Belted Kingfishers are loners (male and female), except for a few short months each year when they pair and work one with another. Their chores together include alternating to dig the long earthen tunnel nest. Then the female lays and begins to brood the eggs. The mate now fishes for sustenance for the pair. They eat small fish, tadpoles, frogs, and aquatic insects. To see the quick and accurate dive of the Kingfisher is surprising and sometimes almost undetected by the eye. However, the splash from the dive and victor's proof of catch in his bill causes confirmation that what you thought you saw was real. When eggs are laid, the brooding is also shared and alternated with more fishing by both. The male broods while the female dines. The female then sits while the male is catching fish. When eggs are hatched, the female warms and comforts the naked chicks while the male feeds the newly hatched. When feathers come, they supply warmth; then the pair works together in finding food. Juveniles are taught to fly and then fish by their parents. They bait the young by dropping dead fish for them to dive for and retrieve. When they have learned to fish, they are driven from the territory by their parents. After this is accomplished the parents each have a long vacation from each other.

When real work, hard work, needs doing, let's do it together. When we pair up on things we can get much more done. Error is avoided when purpose is affixed by a couple together. Alternation of chores keeps us fresh and ready to go when it becomes our turn to help with the home or aid in providing. Our youth are broader, stronger, and more mature when tutored by more than one. Yes, when it's time for young to leave the nest, let's help our children quickly find their way. Remember, its love that is exercised when they are sharply probed along the way. Don't hold on, don't look back. Instead, find happiness in your flights of accomplishment along the way. Oh, yes indeed, it is true; Mom and Dad, as well, need short vacations from each other. Parents, whether together or apart, enjoy success of their young. It is true, absence makes the heart grow fonder and reunion brings pride and happiness to all.

GULL, HEERMANN'S

This small to medium sized gull, 17½ to 20″ long, is sometimes referred to as the White-headed Gull. It has a rounder or plumper appearance than other Pacific Coast Gulls. The adult in summer has a pure white head, blending on its neck to bluish-gray. It is dusk or blackish on the back, wings, and tail, but wing coverts and primaries are fringed with white. The bill is reddish tipped with black, and the legs are black. In the winter the adult is similar, but the head is matted with grayish brown. First year birds' head, back, wings, and under parts are chocolate brown. The tail and feet are black, and the bill is flesh colored with a black tip.

These are in actual fact, Mexico Breeding Birds. They breed mainly on off-shore islands in the Gulf of California and western Baja California, Mexico. They winter on the Pacific Coast from California to British Columbia. They build nests on the ground in colonies with perhaps a few sticks or grasses. Two or three dull drab yellowish eggs marked with spots of brown and lilac are laid. They eat fish, mollusks, crustaceans, as well as handouts from the human visitors. On even the slightest intention of providing handouts of crackers, chips, or anything else men wish to share, the birds seem to come from nowhere. Many tourists come just to feed these birds and some come to picnic but end up sharing with these aggressive beggars. They sing while pilfering; their call is nasal and whining, "*see-whee,"* and also a low pitched "*kuk-kuk-kuk".*

The Heermann's seem to practice reverse migration. After breeding in Mexico, they move north along the Pacific Coast to loaf, to sail in the ocean breezes, and to dream. When food is needed, they either steal from the Brown Pelican (who seem acquired and easy prey) or sneak up on a school of surf's swimming fish from behind. After the somewhat clumsy but successful dives of the Brown Pelican, the Heermann's Gull, it seems, is always there to steal fish flopping out of the pelican's lower mandible. When a group of Herring Gulls are near fishing, these intelligent Heermann's Gulls will carefully approach them to begin to obtain a portion of their catch. The fish school will then dive deeper for protection from the skirmish of both species of birds from above.

Heermann's Gulls lay their eggs on islands during sweltering hot days near the equator and then find it necessary to stand over the eggs like gray umbrellas to protect them from the unbearable heat of the sun.

Life, indeed, needs and enjoys variety and the Heermann's Gull provides some opposites. Instead of going south each winter, they go north. Instead of keeping their eggs and young chicks warm while brooding them, they stand over the eggs and keep them cool. Instead of wearing pastel or light color in summer, they put on a dark coat.

Our first sighting of the Heermann's Gull was in California. One of our favorite places to visit is Point Loma or the Cabrillo National Monument in San Diego, California. There is a road that goes right down to the lower lighthouse beach. Two different families were attempting to picnic, but the wind was blowing and the insistent Heeramann's were having their (the two family's) meals.

Perhaps we should admire their ability to feed. We should relax and not lose our opportunity to be cool, in spite of their unexplainable need for some succor; should we not admire their skill? Perhaps we should not go too far with this life of no concern, though. Didn't Robin Hood take food to nourish the poor? Aren't Herrmann's the poor in "bird-dom"?

PELICAN, AMERICAN WHITE

This species is one of America's very large birds, wing tip to wing tip 9′ with a length of 56 to 70″. The bill is three times the length of the head, is a very bright orange, and develops a horn, plate, or fin in breeding season on the upper mandible. There is a great yellow throat-pouch beneath the lower mandible. This bird is white with wing tips or primary and outer secondary black. As well, at the brooding time of year, a yellow tinge appears on the breast, crest, and on the back of its head. It has yellow legs and webbed feet and never a song to sing, instead it just grunts and groans.

After many amazing sightings, one might describe the great white's circling, gliding flights as astonishing! It is awesome to see pelicans whirling and turning in unison, capable of soaring for long distances - migrations of flight. This sighting seems to flash on, then off; or appear, then disappear as the sunrays reflect off the black primary wing feathers. The sighting is bright when light's angle of incidence reflects to us and disappears when the reflection moves elsewhere. They circle north in spring and south in fall, usually in clockwise rounds. Nature offers few sightings more serene.

They are sublimely handsome in their relocation flights. These birds migrate in wondrous halos on motionless flying wings, using the upward air drafts to stay aloft with its head folded back onto its shoulders to lower wind resistance. Each late March or early April we watch with anticipation the skies, awaiting the joyful experience.

They prefer large inland, shallow bodies of water rich in fish with accessible islands or pinnacles for nesting. They are seen from Canada south to California, and from Yellowstone, Wyoming south to Nevada and Utah. They build large nests on sandy or marsh thickets and nest in colonies near water and high vegetation, laying two or three white eggs. The young are naked at birth and must be constantly protected. Direct sunlight can be dangerous.

The White Pelican never dives from aloft, like the Brown, but instead goes out paddling with webbed feet on the water's surface in schools or alone to fish. A gentle rhythmic motion is maintained while scooping up fish. The head goes into the water, the large lower mandible opens, the catch is made, the head comes up, and the fish are swallowed. Food is carried to chicks in the stomach, not in the bill. After fishing is done, they fly back to the nest. A fish soup is regurgitated where the clamoring young dine from the parent's lower throat-pouch mandible in a rapid frenzy of indulgent dinner.

Another joy in sighting is to watch this great white bird come in to land on the water. Wing primaries separated and stretched out wide, head up from its rest on the shoulders, webbed feet outstretched like small water skis, it scoots across the water then sinks to a stop as breast settles into the lake water.

When participating in a sighting using powerful binoculars, the breadth of our prospective is narrowed, seeing these great white birds together glide, each circle has moved the flock closer to its goal. Life is full of circles or progression. Each week, each month, each year, we compile experiences of both mundane chores and delightful incidents. In our latter years, we experience joy at remembering and sharing life's many lessons with those young men and women about us. Let us remember that what we are "bringing up" to share with them is not always wisdom from the rounds in life, but "puke" instead! Enjoy the uplifts of life. Find comfort in the telling of the revolutions or progressions you have enjoyed; but perhaps we should tell the truth, shouldn't we? -Not just tall tales!

WARBLER, YELLOW

An exceedingly beautiful little burst of sunshine, the Yellow Warbler is 4½ to 5¼″ long. This bird is almost always sighted beneath a canopy of green foliage. One of the most widespread of warblers in North America, the Yellow Warbler has a plain or blank-looking face where a distinct, beady black eye appears. The bill is shorter than the head, the tail is tiny, and rounded wings are long and pointed. The male face and underparts are yellow with bold chestnut streaks underneath. Females are duller than males and yellow is sometimes blended with olive-green. Both have a lighter yellow eye ring and yellow wing bars.

They sing a lovely cheerful titillation song which seems to say, "*Sweet, sweet, sweet - I'm so sweet*". To view this elusive little bird, you must start by looking toward its song, then gaze toward the rustling of leaves; then sight a quick yellow streak of flight, hopefully to a new perch where a short view can be obtained before this colorful creature is off again. I have lain on my back looking up at the song, *"Sweet, sweet, sweet - I'm so sweet,"* for many minutes before catching a good glimpse of this brightly colored song bird. I found a picnic table once and reclined for over an hour looking up in the cottonwood trees at the Yellow Warblers. Once you have found this yellow warbler, pictures can be snapped of its brisk full life, which is indeed full of melody. Biological necessities of songs being sung for purposes of bird life are obvious: mating, territorial defense, the cheerful announcement to a flock of finding food, or to alert of danger. However, to hear the Yellow Warbler sing causes birders to wonder if they don't just sing for the pleasure of singing.

A deep cup of silky plant down, grasses, and spider webs form the nest, which is attached to a crotch of new tree, sapling, or shrub growth. Three to five pale whitish eggs spotted with brown around the large end are laid.

The load of life's responsibilities seems often wanted to be given away by other birds. The task of brooding, feeding, and raising young seems to be a battleground between the lazy Cowbird and the industrious Yellow Warbler. The Cowbird will drop this burden of eggs to brood and chicks to feed into many non-suspecting nest builders each summer with the Yellow Warbler as a main target. The chore of providing insects and spiders for family sustenance now has become immense.

There are many predicaments in life where it is indeed hard to decide just what to do. May we learn to weigh our options carefully, like the Yellow Warbler! When the Cowbird drops an egg in the warbler's nest, still without a warbler's egg in it or only one, the Yellow Warbler avoids the burden given by another by abandoning the nest or by disregarding both her own egg, as well as the Cowbird's egg by covering them with additional nest. However, when she has two or more of her own eggs in the nest and has started to brood them, the additional burden is accepted. Perhaps her young can beat the short incubation time of the Cowbird and with more equal size chicks in the nest compete more fairly for the never-ending supply of food provided by the warblers. When tough decisions come our way, may we learn to decide the best course to take! Let's move forward and bear the load or start anew with a lighter burden and lost time to achieve. A pair of Yellow Warblers makes their decision quickly and with a never ending song in their heart, victory becomes their attitude, causing new young to fly very soon.

EGRET, SNOWY

This fine structured graceful egret is a medium sized heron, 22 to 26″ long. The wingspan is 3¼′ in length. It is an elegant plumed bird. Aigrette crown feathers turn upward on breeding adults and the throat and breast feathers train low like lace. The sexes each look alike. It gives a low croak sound *"aah-raarrh"* as its song. The Snowy Egret has dazzling white plumage on its head, neck, body, tail, and wings. The bill is slender and thin, black, and quite sharp. Its eyes and lores are yellow, and it has long black legs with bright yellow feet. The immature have some yellow on the back of their legs. The Snowy is easily confused with the immature Little Blue Heron. Cautious birders, however, will notice the difference; for the Little Blue has all yellow legs and feet.

Snowy Egret are widespread in North and South America and are found in most wetlands, ponds, shorelines, swamps, marshes, lagoons, and wet meadows. They are found in the southwest, mid-west, the south, and in Central and South America. Two to five pale blue-green eggs are laid in a platform of sticks on the ground or in brush or trees 5 to 30′ above ground.

The Snowy, even more than the Great Egret, was sought in the late 1800's and early 1900's for its delicate plumage. Women's hat and dress styles almost destroyed these decorative white birds. Its ability to adapt and expand its range in the last few decades gives us the chance to still admire these great white birds.

The plumes or aigrette feathers, although shorter than Great Egrets and Herons, are more delicate and ethereal (airy or heavenly in appearance) than any of the herons.

The Snowy are more aggressive foragers and feeders than other herons and egrets. They hunt briskly in muddy waters. They partly extend their wings to cut out glare from the sun. They balance on one foot while raking the other yellow foot to scare up the prey. Their desire is to find small fish, frogs, snakes, insects, crabs, and other shellfish.

This white egret's display consists of bowing with its neck extended and the ethereal aigrette feathers raised. This includes bill clicking, which draws attention to their stylish white dress and golden slippers. The male holds out a branch in his bill and the female takes it, as if to say, "I do," much like we humans do. Yes, the beautiful Snowy Egret should remind us of beautiful occasions, sacred or heavenly times in our lives where fine lace, smart dress, and commitment cause us to put our best foot forward. Yes, with perhaps golden slipper feet, like the Snowy Egret. To dress, trim, groom, adorn, embellish, or arrange always makes life more meaningful. The hours spent sewing or creating with material and thread surely should be much appreciated. The time spent scripting, arranging, or planning personal events or parts of life indeed build memories and commitment or purpose to our future. To find dignity, show lace and beauty in nature as the beautiful white Snowy Egret does, should remind us of the purpose of those special occasions and what all the work and preparation was for; such as in a civic, fraternal, or community affair or in a religious ceremony of promise, devotion or edification; as well the wedding, where men dress sharp, clean and distinguished with a boutonniere for color and a shine on the shoes; the bride's hair groomed, curled, and combed to perfection, dress sparkling with lace and pearls with veil and train shimmering with flare and bounce. What a beautiful sight! What a display of devotion and promise! Let the sightings of the Snowy remind us of those times of pledging.

GULL, RING-BILLED

This is a most widespread and recognizable seagull. The Ring-billed gull is a small-size gull, 18 to 20″ long, with a wing span of just 3¾′. It has a sloping head and medium-size bill. An adult's plumage is acquired in about three years. The adult is pure white with a pale bluish-gray mantle. The bill is yellow and its song is a "*kree, kree, kree,*" and occasionally it will squeal a high shrill, "*ky-eow*". The upper and lower mandibles have black coloring near the tip, creating the appearance of a black ring around the bill. In adults, the eyes become pale yellow, but they are darker in juveniles, and each has a reddish eye ring. Legs are quite bright yellow-green. The head is found streaked with brown in winter. First winter birds have pinkish mandibles but look as though they were dipped in black ink at the tip. The head is light gray and has light brown bars. The breast is similar, but the flanks are a few shades darker in their brown and gray markings. Second winter birds no longer have black marks on the bill at the tip; they have moved closer to the head and have created a quite broad ring-bill effect. The face, neck, and breast are white, but the back, crown of head, and nap of the neck are flaked with gray. One of this bird's distinctions is a rapid wing beat and a rather buoyant flight. After becoming familiar with this style of flight, it can be identified from great distances. This bird is easily confused with the Herring Gull. On land or beach, a very brisk stroll (almost a walking race) against other Ring-bills exists as they forage for food. Some keys differences exist which help to know that the sighting is that of a Ring-bill, not a Herring Gull. The head and neck are more fragile and less pronounced than the Herring. The Ring-bill has larger white spots on its primaries, more extensive black undersurfaces of its primaries, and a reddish eye ring. Of course, the ring around the bill is the real key.

The Ring-bill breeds on islands near coasts or inland lakes in the west coast, upper Great Basin, and east over large portions of the United States and Canada. Ring-bills usually nest in colonies along with California Seagulls and lay about three buff colored eggs in a nest built of grasses and sticks on the ground.

Like many other gulls, Ring-bills eat worms, insects, grubs, and similar fauna. When gulls swarm near restaurant refuse areas, there are usually Ring-bills and California Gulls that are there sorting through the trash for edibles. Ring-bills, as well as other gulls, raid nests of sea birds for food. A most delightful sighting of Ring-bill Gulls was experienced in Franklin County, Idaho. These pleasurable gulls were feeding behind a farmer as he furrowed his field in preparation for spring planting. They would flutter here and there behind the plow. These flights to find the right spot were followed by their unique brisk saunter (race-like walk), each wanting to be next to capture an earthworm or grub or some kind of meal. We, as well, are encouraged to lengthen our stride, to step forward by being exuberant, showing interest in each day. Would we not accomplish more if we could approach this endeavor with the enthusiastic, competitive vigor shown in the fluttering flights and spirited walks of the Ring-bill? We would not only accomplish more, but would delight in the endeavor, as well. The aggressive, brisk, and diligent walk of the Ring-bill not only provides exercise for the body, but it also causes an attitude change in what we need to do. Do you not find that a good brisk start brings victory? So increase the length of your pace for success, as does the Ring-billed Gull.

SWAN, TRUMPETER

This is a swan of forest habitats. The family in the painting above was first captured on film north of Jackson Hole, Wyoming. Yellowstone, The Grand Teton National Parks, and the nearby mountain lakes all provide nice sightings of this great water bird. In fact, the mountain gravel road that runs east from Ashton, Idaho, to US 89 - 287 has a Swan Lake where Trumpeter's nest annually. This bird is the largest of all North American waterfowl, 60 to 72″ long (up to 28 pounds) with a wing span of nearly 10′ long. Its muscular neck is extended in flight and held straight up when swimming. Its forehead and bill profile are long and flat, giving it a well displayed look. His visible head and neck are never hidden because he is the king of waterfowl on his chosen lake.

Its plumage is pure white, and the bill and feet are black. Pink can be seen on the outer edge of the mandibles. The young are shadowed white, and the bill shows much pinker. The cry of this bird is a deep resounding trumpeting noise, *"Koo-hah" or "Ko-bob".*

Birding history would report this wondrous bird as gathering in mighty flocks, coming full circle into the sky affixing for their return. The slaughter then began: pelts were turned into powder puffs, down for pillows, and feathers for quills and adornments. Finally, after having been reduced to a few hundred birds, attempts were begun to save them. As these handsome birds display and pair for life, the male shows love by caressing his mate with a fond petting motion of head and neck. The wooing shows at times male and female with necks intertwined together in this beautiful ritual of courtship. They do not pair and breed until four years of age. Perhaps this maturity is the cause of their perfection in parenting. The male is the strong sentinel and guard. Young leave the nest as they hatch and the female is the key in nurturing them. These cygnets are carefully taught how to forage for food. This bird cannot dive for food but does bob or claw for food at water's bottom; they eat aquatic plants, roots and insects in fresh waters. As well, they are taught to protect themselves and finally how to fly.

It seems these days that the Trumpeter both breeds and lives in wetlands which are protected, such as the national park previously mentioned. Once they could be found breeding from coast to coast and from deep north into Canada and Alaska, to the northern United States. Protection has brought the Trumpeter back to nearly five thousand.

Both male and female combine to build a large nest. The nest is constructed of grasses, rushes, feathers, and down on an island or small peninsula in dense fresh water rushes. They lay three to nine dull white eggs. Cygnets separate from parents after one year.

While my wife and I were in Jackson Hole, Wyoming, we had the opportunity to observe a Trumpeter pair and two young enjoying the sweet summer sun and the rich green wetlands of the National Elk Reservation. The male, indeed, was on lookout, and as Canadian Geese came too close, an attack was begun: wings flapping, webbed feet slapping the water, and the intruding geese honking with fear scampered away. What a master of domain this great bird is! If we could all meet danger as this dominant male did. A beautiful home scene: this Trumpeter family enjoying the rich surroundings was really possible because defenses were in place. The pleasantries of life, and family, are indeed worth protecting. If we were always ready to guard like this great monarch, fewer would challenge!

DIPPER, AMERICAN

Few birds are more specialized than the extremely aquatic American Dipper. This bird is robin-size but looks smaller because of its short up-turned tail. It looks like a small dark tea pitcher, 7 to 8½″ long. Its plumage is dense and soft. The Dipper is kept warm and dry by special oil glands as they live near, around, or under the clear cold waters of Rocky Mountains streams. Its color is dark slate gray or wet black, and they have long wiry legs and claws to grab the underwater rocks, roots, and soil.

This tireless worker sings variable, loud, and musical songs virtually all year long, always happy in sunshine, rain, or snow. The Dipper is at all times vibrant, busy, and on the go. Its call is a sharp "*zeet,*" sounded above rustling waters.

Dippers are found from Alaska to Central America, and from California to the east slopes of the Rocky Mountains. This American water ouzel finds foaming, cold clear water springs and creeks to call its home.

This dark small bird nests near or under a waterfall, on a root shelf, or earthen protrusion. A dome-like hut is built of living green moss. There is a door on one side, and from three to six white eggs are laid for the mother ouzel to brood. If the live green moss nest is not watered by the spray of the falls, the male Dipper will sprinkle it with his wings. The male performs this chore as he works to find and serve his brooding female and her eggs, then chicks.

With short and rounded wings, sharp claws, large feet and strong legs he swims, walks, and firmly stands against the stiff currents of the cool mountain streams. With his head down and pointed upstream, he forages for food—probing under rocks and roots, seeking flatworms, snails, larva, rock rollers, and sometimes snatching a small fish. When food is obtained the flight home is performed while following the stream—darting back and forth, going with each turn through every narrow very near the bubbling water upstream, through the falls to serve his hungry family.

My immediate attraction to this bird came from its desire to find a prominent point adjacent to a mountain stream. While perched there, he performs his lovely dance. Thus, a series of prompt and distinct deep-knee bends are rhythmically carried out. This distinct dipping motion, always while perched, is of course, the cause of his name, the Dipper. At Boy Scout Camp Loll, they dance in front of Polar Bear Spring; at Camp Kiesel, they fly swiftly above the wheat grass's blissful small brooks.

Should we not try to emulate the Dipper in our lives? Gripping tightly to the truths of life we can face the strong, cold currents of opposition and contention, always with a song and the vigor of forward flight. Repeatedly using the lush foliage and flowering creek banks, the curves of the stream, and perhaps a waterfall to stay close to things precious and beautiful, but protected from the predators of the world. At all times serving, working, being active and vital, using the swift, clean, and constant forward motion of the mountain waters as his example to keep moving forward; yet knowing when it's time to stop and have the joy of a little dance. Let's know those good traits and be more like the Dipper in our lives.

CARDINAL, NORTHERN

This stunning and highly successful American Finch is one of the most recognizable birds in North America. The male cardinal is 7½ to 9″ long. It has a stunning peaked, tufted black color that masks the face above with a distinctively large reddish-orange stout conical bill below. The cardinal wings are short and rounded; and the tail is long. The female bill is more pinkish than red. The male is bright red, but the crest and wings are reddish-black. The female is olive-buff with buff-reddish-orange crest and wings. The adult birds usually appear quite plump. Juveniles lack the distinctive face mask. The bill is gray and the male, although browner, resembles the female (reddish tones are yet to come).

A distinctive high, but whistled "*chee-chee-chee," followed* by a "*weight-woight-woight-woight,"* is uttered in sharp notes, and both sexes sing most all year long.

The cardinal is found in suburbs, as well as city parks, gardens, mesquite, thickets near water, and mixed woods. Because of their bright color, they seek sufficient cover for this vivid species' comfort. They are mainly found in the states south of the Great Lakes. But in the West, they are found in the Colorado River Valley of California, Baja California of Mexico, and also introduced in Hawaii.

The cardinal, having a stout bill, is a seed, fruit, and insect eating bird and is attracted to feeders easily, but it mainly feeds on the ground. These birds are not migratory, thus they desire food to be plentiful in the surroundings or at a bird feeder close by. Inclement snow or harsh weather dos not seem to drive them off. Consequently, they seem to like mild winters the best.

Disregarding their striking color, they attempt to be quite elusive, desiring privacy by hiding in thickets. However, the male desires to sing and finds an appropriate limb (on stage it would seem) where he vocalizes his songs. These colorful birds stick together for life. Their devotion to each other is strong through all four seasons. However, each spring the male resumes his courtship skills. The female will brood more than one clutch annually, and each litter provides the same manner of care from the male. They are ever happy and vigorous and always productive. What accord is found in their lifetime commitment!

Upon returning to the same nesting area (April-June), all of the following are usually done in a thick bunch of leaves 2 to 12′ above ground where a loosely constructed cup shaped nest is made. Three to five pale green eggs with light brown-lilac specks are laid. Many times the male continues to feed the young of the first clutch while the female starts anew with another nest and brood.

In a strange way, the cardinal pair cooperates while the nest is being built; the female builds the nest with the male close beside on a perch singing delightfully to her while she works. Of course their territory must be defined. He is like the property owner and his refrains in song protect the potential harvest required to feed his family. In spite of his lack of labor, his attention or fondness for his mate is clearly noticeable. Ornithologists will tell that his voice yields protection from intruders, but could it not be that he sings for his kin? She broods the young as the male provides seeds to his mate and their chicks (bill tip to bill tip) during this incubate period. His duties take him away from the nest, but somehow his concern for mother and children stays affixed. He reveals his love and she responds with a soft call in return. What a family to admire!

GNATCATCHER, BLUE-GRAY

This active slender billed, tiny, and lean songbird is about 4 to 5″ long. The tail is slim and long, rounded at the tip (nearly one third to half of its length). Some would say it resembles a miniature mocking bird.

Gnatcatchers obviously eat gnats and other insects, including spiders. They gather from tree and shrub trunks, branches, buds, and leaves. This gathering process provides the birdwatcher with quite a delightful display of movement in flight. They snatch gnats in mid-flight or soar above blossoms to nab unsuspecting insects. Their diet is almost exclusively insects, both the leaf-eating kind and those that nest in flowers. These insects are considered pests by man.

As the Blue-Gray Gnatcatcher feeds, it makes a thin boozy insect-like vocalization, "*speee*" or "*pseee, tsee,*" followed by more warbled or jumbled phrases. Perhaps it attempts to sound like an insect to lure them to its grasp.

The upperparts are blue-gray and the underparts are white to grayish-white. The secondary wing feathers have white edges. The tail is white under and black above. The white outer feathers are seen from above. It has a thin white eye ring.

The male's crown and brows (mainly in summer) are bordered with black. The female lacks black markings and is less bluish. Playful rapid movements of the tail reveal flashes of black and white.

Gnatcatchers commonly carry their tail cocked but twitching and angled to their body or held straight back and down, usually slightly fanned. It is a very active, fidgety little explorer constantly darting, hovering, rising, and diving in pursuit of its elusive prey. Despite their frail appearance, gnatcatchers are not afraid of anything. They will show aggression toward intruders, such as Blue Jays and Crows. Because of this, insignificant care is taken to hide their nests. As a result, predacious birds do lay their eggs in their nests.

This bird finds home in a variety of forests, such as deciduous hardwood groves of trees and shrubs in the north, east, and south, to chaparral forests of Scrub Oak, Juniper, and Pinion Pine in the Midwest, and west from northern California, east to southwest Colorado, and from southern Ontario and New York south to Central America. It winters in southern states and mid-America. The nest is a compact cup of grass, bark strips, and other fibers slung between twigs in a tree or shrub from two to seventy feet above ground. Three to six eggs, pale blue with reddish-brown blotches are laid. The female lays her eggs, incubates them, and broods young. The male does take his turn in incubation, but his primary role is to alarm intruders to stay away by continually singing his buzzing song.

These delicately framed winged creatures choose woodlands which are so beautiful that they seem sacred. There are always some forest neighbors we might call predacious, for they wish not the chores of brooding, feeding, nurturing, and caring for their youthful chicks. These birds lay their eggs in the nests of others. The pair of Blue-Gray Gnatcatchers slacks not at all from eggs needing warmth to birds practicing and ready to fly. There are many in life who care for all, all the way, like the Blue-Gray Gnatcatchers. My son's scoutmaster taught and trained him, as well as others. His own son was not favored. My daughter's leader loved her and helped her and others design for the future, just as she did her own daughter. We should emulate those great youth leaders and support them in every way we can. We, as the gnatcatcher, care for all - no exceptions!

RAVEN, COMMON

The Common Raven is 22 to 27″ long. Its wings are long and pointed. The primaries look like fingers when soaring, and its wedged shaped tail is much shorter than its wings. The bill is large and has a high ridge on it, making it higher than broad. The upper mandible is mustached with glossy feathers. Its head is large, neck is thick, and throat is bibbed with shaggy ruff feathers. The raven is the most imposing of birds in the crow family.

Its deep, harsh voice cannot be mistaken. It utters a croaking "*kraak,*" or sometimes a melodious "*Kloo-Klok,*" usually during flight. It differs greatly from the "*caw*" sound of others in the crow family. There are a number of other kinds of notes or noises which have been recorded around its partner or chicks. These sounds of commotion are quite demonstrative for a large black bird.

Its entire plumage is black. Wing covert, secondary, primaries, and tail feathers show a glossy violet metallic sheen. The underparts' feathers show blue or greenish blue in their sheen.

Common Ravens scavenge along highways for refuse in many environments:

deserts, mountains, canyons, forests, Pacific beaches and cliffs. They have been seen north to Oregon, Montana, South Dakota, and in parts of Canada; south to Honduras and east to the Mississippi River, and in some states just beyond. Ravens will come into almost any surroundings to satisfy their quest for food.

Nests are almost always located on a ledge or on an inaccessible cliff. The territory sometimes causes them to choose a tree, but not often. Nests are composed of a large interlaced stick base, bones, soft fibers, coarse grasses, and bark strips. Four to seven greenish, heavily spotted with brown eggs are laid. Incubation takes 18-29 days by the female. Young leave the nest in six to seven weeks.

These birds are omnivorous and seem to always be around where mankind strews or throws his garbage along roadways, near national parks, hotels, and eating establishments. They are also fond of carrion; thus assisting in the constant service of cleaning up road kill.

A raven has been associated with divine mystical and magical powers and with messages of assistance. Common Ravens cause great surprise in flight, showing aerial acrobatics of tumbling, rolling, soaring, and even racing or chasing hawks and eagles.

This large black bird seems capable of making good life happen everywhere. Wherever this raven decides to wander, its cunning, ingenuity, and intelligence seems to make things work. The reputation of the Common Raven stands high in each environment. A very sharp and perceptive

bird, it seems to apply reason in situations entirely new to it. It shows insight equal to or greater than man's best friend, the dog. Native Americans respect the raven's cunning and ability to survive and depict their attributes in ceremony and dance. As we are all from time to time blessed with new neighbors in our lives, let us remember the skill of the raven as it expands its territory. They love their partner and their offspring. They are intelligent. Let's not judge by appearance or mistake inquisitiveness or wanting to learn for snoopiness. Let's really make our new neighbors our friends. They may be as intelligent and resourceful as the raven!

TERN, ROYAL

This is a large tern, 18 to 20˝ long, with color patterns similar to gulls. They fly with the bill turned down, have long pointed wings and a forked tail and have a cap and crest on the back of the head. During display, the cap seems to come around the eye. The tips of the wings reach almost to the tip of the tail. The feet are small and the legs are short, making it a very poor swimmer; however, it is tireless and diligent in flight.

It makes a much higher pitch sound than other terns, a long "*chirrup, kee-rer*" escapes its throat, but not often.

They are mainly white, including the underparts. The breeding adult has a glossy black-greenish cap and a rough crest. When not breeding, the forehead is white. The mantle is blue-gray and the primary feathers are dark on the edges near the tips. The bill is long and orange, tinged with red. The feet are black, but in some instances they are orange. Juveniles have white foreheads and yellow bills.

Their diet consists almost exclusively of fish. Adults go out to sea and catch small fish and sometimes squid by diving from heights of 40 to 50 feet. However, they will steal from other birds, like pelicans, whenever they can.

Royal Terns nest in large colonies on the ground (a mere depression in the sand). Some of the islands they use will have as many as 10,000 to 12,000 nests. No other bird in North America nests so close together, almost touching each other but not quite. They lay two to three pale buff or whitish eggs.

They turn in flight while searching for food, carrying their long bills pointing downward (a distinguishing feature in flight); thus, ready to dive for food. They have the unique ability to stop just above the surface of the water to survey for food. If none is spied, they climb again to repeat the whole process.

It seems that somehow almost all eggs become hatchlings within a very few days, which ensures the likelihood of most surviving.

These young downy colonies run about on the sand before they can fly. Adult birds shepherd these extraordinary crèches. They can be led around as a unit and caused to hide from marauders or to just get out of the sun. Crèches can be very large, even thousands of little birds. Crèches are usually quite helpful. Even though they can make it difficult when young birds want to be fed, for each must make its call heard (each chick trying to "out-screech" the others) and move to the edge of the crèche to meet a parent who amazingly will only feed its own chicks and no one else's.

Many things in life are better accomplished in groups. In business management, decisions are usually more correct and successful if a number of people (Mastermind Theory) are involved in the determination. This also wins support or conversion to the plan. Passive resistance has little or no effect unless great groups are gathered to cause force. We elect public officials by ganging together to amass a great vote for them. Is not our faith strengthened by singing and worshiping together? There is great strength and protection when birds form crèches.

Yet, their parents find each individual - to feed and nurture the one. Let's get in concert. We can and should pattern our lives toward the righteous purposes found displayed by the amazing young Royal Terns. We can demonstrate togetherness to accomplish many things of value and importance for ourselves and others.

GOOSE, CANADIAN

Ornithologists suggest there are about eleven sub-species in the Canadian Goose family. There are basically three: The Greater: 32 to 48″ in length with a wingspan of as much as 5 long′ and a weight of 13 to 14 pounds. The Lesser Race: 22 to 27″ in length with a short neck and a weight of nearly 9 pounds. The Cackling Goose: 23 to 25″ in length with a very short neck and a weight of 2½ to 3 pounds.

Their calls are essentially two - Small Race, known as "Crackling Geese," gives a high pitched or crackling sound "*yelk, yelk a lick, a lick or lick".* Larger Race, Known as "Honkers," who give more drawn out honking calls: a deep, musical "*honk-a-lonk".*

All have the same basic color distinction with some coloration variance between gray and brown in the Lesser Race. Its bill, legs, feet, head, and neck are black. It has a white chinstrap from ear to ear with a white U-shaped rump band, belly, and flanks.

Canadians feed in shallow waters by "tipping" for aquatic plants and small aquatic animals and also by grazing pastures and crop fields for grass and grain.

Its habitat consists of varied marshes, lakes, rivers, tundra, forest bogs, and prairie sloughs. When people feed them, such as a golf course putting out cracked corn to feed the birds, the flocks become semi-domesticated and stay in the region year round.

Mated pairs stay together usually for life. The young remain with their parents until they return to the nesting grounds in the spring. Four to eight white eggs which vary according to the Canadian goose species. In wildlife refuges, platforms are built to stand above shallow water where straw bales are placed for geese to build nests and line them with down.

The Canadian goose is a well-known goose and one of the most common sightings is when they "V-up" to migrate. Part of the joy of these sighting is their conversation as they honk to one another and announce to us bird watchers of their approach. Shortly after the young hatch, the pair begins to molt. Nature has chosen this time because they must stay close to their little goslings. It is a dangerous time because of the loss of flight. The Canadian goose can often times be sighted when feeding by the distinct high neck and head of the gander, their chosen sentinels or guards.

The care and keeping of the little goslings is always fascinating to watch. We show in the painting the gander, head held high, always on the lookout as they proceed with the young between the pair. Note the goose's bent or kinked neck, as her head hovers over the young, ready to signal them by voice or head motions for the need to line up for group movement; spread out a little and forage...there is good food here; it's time to hide...come over here in this tall grass; or if swimming...to alert to dive and swim away.

Little goslings can swim under water up to nearly 5 yards when alerted to do so. Of course, there is always a constant parent to give the youth directions. We should approach parenting like the Canadian goose - with order, discipline, care, and tender love.

OSPREY

This is a hawk-sized, hook-billed Fish Eagle, 24 to 25″ long with a wing span of 54 to 70″, tip to tip. Looking up, the inexperienced birder might mistake this bird for a gull; however, the long whitish color under the wings shows a crook in their wings and brownish-black marks at the wrist. Of course, when they stop their soaring and start to hover the sighting is known as an Osprey.

Its call can be quite loud, especially when aroused, a high whistling, "*kip kip kip k-yews*". During mating their voices are mellower soft whistles and chirping calls.

Its coloring is brown to brownish-black above, and white below with buff heart-shaped spots on the breast feathers. The head is dark with a white crown and dark eye streaks which extend to the nape of the neck. The bill is bluish-black and the feet are grayish-blue. The talons are black and the tail barred.

It feeds exclusively on freshly caught fish. The young, after learning to fly, need a few months before they are able to sustain themselves as great fishers from the sky.

They breed in Canada, the United States, Europe, and Asia and winter in southern coastal areas.

Ospreys nest in trees or sometimes on the ground where cliffs provide an outlook. Where power poles intrude their territory, usually the power or phone companies will provide platforms on top of their poles for Osprey nesting. Two to four yellowish or buff-white spotted or blotched at the large end with reddish-brown eggs are laid. In large stick nests, Ospreys, like Bald Eagles, return to the same nest year after year, always adding new sticks to riffraff; thus causing the nest to continue to grow larger. Many times nests are a considerable distance from their fishing waters, causing long anxious times for the female to wait as expressed in the painting above,

Ospreys, like other raptors, use effectively the skill of soaring on natural up-drafts. Ospreys have been measured achieving amazing speeds. When prey is observed, a quick study is required before committing to dive. They spy quickly while hovering and then begin their descent. This bird is well equipped to catch unexpected fish with four razor sharp talons curved to full half circles. One toe can reverse so that it has two claws forward and two claws back, making a clutch most powerful and deadly. The Osprey has been known to seize prey too large for them to handle; however, the osprey has powerful wings and tail along with the needed muscular energy to succeed. Power and skill usually cause it to fly away victorious with food. The Osprey always flies to its nest carrying the fish head forward to aid the flight. Ospreys are smaller than Bald Eagles and are often bullied by them, which usually causes fish to be dropped, allowing the eagle to steal the fish.

Many times we have watched the Ospreys south of Jackson, near a town called Alpine in West Wyoming. The power workers have placed wooden pallets on the top of power poles. The Ospreys then build nests on these convenient platforms. The female is most interested in the catch her mate is bringing to her and their anxious chicks. How important is a good provider, one who takes his responsibility seriously? As we toil each day, let us remember those we love and why we work hard. The labor is for home and family, not personal aggrandizement. See the pleasure of the female Osprey; she shows why we work hard!

FALCON, PEREGRINE

The Peregrine Falcon is a magnificent and powerfully built raptor. This is a medium-size falcon, 15 to 20″, with long pointed wings that span from 39 to 45″ tip to tip. This bird also is known as the Duck Hawk and has a stern look caused by the skull bone eye shield. The bill is sharply hooked, toothed, and notched.

The tail is quite short with curved, long, and very sharp talons. The female is noticeably larger than the male. They are usually silent, but during breeding or when excited, a rather harsh "*kek-kek-kek-keek*" sound is made usually in close proximity to the raptors' nests.

It has a dark crown, broad dark sideburns, bluish gray-brown back closely barred, and a white softly spotted bib. The breast and belly are cream colored and finely barred.

It breeds on cliffs in coastal areas, mountain regions, and on buildings in cities from April through June. Its habitat is from north of the tree line in Alaska and Canada, south to mid-Canada, Pacific Northwest, and south to the Rocky Mountains and Mexico. It is also found worldwide in Europe, Asia, Australia, Argentina, and Chile.

Preying almost entirely on birds, they are known to show great courage when food is desired by attacking birds much larger than themselves. They prefer waterfowl in open ranges and pigeons in the cities.

Eggs are laid on bare rock, hard stone, or brick. Both sexes share the brooding of three to five eggs, which are cream colored with reddish-brown marks. Because of their desire to prey on pigeons, the sill of a hotel window works fine for a nest.

In spite of the fact that the Peregrine Falcon was the first bird put on the U.S. endangered species list, it is one of the most numerous of the birds of prey. This bird is truly one of the amazing predators, very specially equipped (slender, fast, and powerful) to easily take on birds of great size. Like most predators, they rely on sharp talons to make their kill. The beak is designed for cropping flesh or grinding bones. Although the adult bird knows how to hit hard and stun or kill, the youngsters, not yet skilled in the dive may fail to stun or take their victim. The parent falcons begin to make the skill of providing more important to the young by failing to make available fowl to eat. It has no superior among birds. Its ability to fly as a destroyer is real. Built like a fighter plane's bullet head with broad shoulders, shaded eyes, and retinas nearly six times more sensitive than that of birds they favor to quarry upon. The Peregrine is superbly sleek, capable of dropping into a power dive of 175 to 200 miles per hour. These dramatic features are better used in the field for identification than the normal color and feather markings. Man's association with this bird is to use its hunting skills (falconry), which has now become a sport.

No one likes a fight, but in this world conflict seems somewhat unavoidable. As the falcon, should we not quickly circle the situation, thus evaluate our prospective that our view of the potential conflict might be directed from on high? When the assignment is identified, set upon the solution quickly with the power and skills which have been made available to you. Victory is sweet but must be followed by compassion and helpfulness. Life most often has been made better by the conquests of suppressive forces. U.S. Air Force Academy cadets cry, "We are the Falcons!" and we now know why,

GULL, FRANKLIN'S

This is a small compacted gull, 13 to 15″ long. Gulls are assumed to be commonly found on the coast, but this bird is rarely, if ever, seen on the Atlantic or Pacific coast. This is a bird of the prairies. In fact, the farmers of the Plains States and provinces refer to it as the Prairie Pigeon. Like most gulls, the Franklin is web-footed and swims well, but its skill in diving is absent.

Its song is a high-pitched "*weeh-a weeh-e weeh-a,*" followed by a soft "*krrr-uk*". When breeding, a "*po-le, po-le*" is heard.

The breeding adult has a black hooded head with a broad white broken eye crescent. Its bill has a reddish tinge with a dark red-black mark near the tip. The back and wings are blue gray and dark gray. Its blackish-grayish wing tips are crossed with white bands. It has white underparts. From September to May, its head is half hooded with gray. Its legs are dusk red and its tail is grayish in the center, unlike any other gull. It's the first winter bird, having a bold white eye crescent, half hooded head, dark wing tips, and a broad black tail band. The juvenile is clear cut with its blackish half-hooded head, bold white eye crescent, and tail banded with white. There is a white thick broken crescent around the eye in all seasons and ages.

Inland marshes, lakes, and beaches are the habitats for this bird. It is commonly seen on farmlands following plowing or scavenging after irrigation. They are found from Alberta to Manitoba in the north and from Oregon, Utah, and Montana to Iowa in the south. They winter in the Gulf Coast and South America.

Large colonies nest on the ground or on floating platforms made of dried reeds in lush growing marshy areas. Three buffish eggs with brown blotches are hatched out by both sexes in about 18 to 20 days. Nests are conveniently spaced, and young downy puffballs swim through the marshlands among the rushes almost immediately after breaking out of the eggs.

This gull is a major consumer of agricultural pests and is considered a very beneficial bird. It consumes nymphs, dragon flies, grubs, cut worms, grasshoppers, or any insect or larva available.

The Franklin, again like most gulls, shows gracefulness and ease in flight. They are clever and accomplished as they maneuver in sometimes stiff breezes against other Franklin's catching food behind a plow or hawking in flight. These delightful acrobatics and competitive flights grant the bird watcher memorable sightings. When feeding is completed, they begin their return in lines or formations and are often seen as they return to rest. Sometimes the Franklin has been referred to as the "Rose Gull" when the underpart feathers are opened up and a faint rosy color is revealed. In summer, and especially during breeding, the breast becomes rosy and the hood gets even darker black.

Life is full of warm feelings. In each of us glows the warm light of love or charity. Aren't there times when people seem to radiate? For example, the glow or blush of young love, the radiance of the face of a mother to be, or the compassion of a prophet or preacher of amorousness. We should awaken our affection properly at appropriate times and show the rosy warm gleam within ourselves. A little more charity will lighten the load for all of us; and all we give seems to be returned with care.

DOVE, MORNING

This medium-size dove has a rounded head, slim neck, and an elongated body that measures about 10 to 12" long from its beak to the end of its tapered and pointed tail. Its tail, consisting of relatively narrow feathers, is longer than its wings. In flight, the two wings show a bend at the wrist. Its dainty movements (when perched) merge with this dove's melodic voice.

Head, chin, and breast are gray-brown washed with glossed pinkish-buff or cinnamon. The bill is dark. The eye ring is light blue, and the lower base of the eye patch has a black spot. The wing coverts have round black spots, and the rump is darker gray-brown than the back of the body. Underparts are softer in color; legs and feet are light rose. The long tail is gray-brown, and in flight shows the feather tips on the outer edges narrowly bordered with white.

The Morning Dove is one of the most widespread birds which are native to North America; its flight is swift and direct, no coasting. Their wings create a quite distinct fluttering, whistling sound, especially during the mating and nesting season, which is a distinguishing feature of this species. This dove can often be seen in large flocks, especially in agricultural areas. As the male courts a prospective mate, he displays a climbing flight, which comes to an end in a series of circular glides on fixed extended wings.

They eat mostly seeds but will occasionally eat insects. They usually feed in flocks, but are often seen at bird feeders singly. Their offspring, like many other doves and pigeons, enjoy a diet rich in fat and protein. A light colored substance is referred to as pigeon milk, which is not milk at all but a liquid matter secreted by glands in their crop. This nutritious food is obtained when the chick thrusts its head well into the parent's very wide open mouth.

The Morning Dove is found in Alaska, Southern Canada, the United States, and Mexico, but is not found near dense forests or wetlands.

The nests they build to hatch out eggs and comfort chicks is indeed flimsy in structure; twigs barely balance on a low tree branch or cactus on the ground and are so fragile that eggs quite often spill from the nest to their destruction. Morning Doves are one of the first to arrive in spring. Breeding starts early and birds may nest one to three times or until mid-September. They are not gregarious and nest alone. One or two white eggs are laid and are incubated by both sexes for about 15 days. The young achieve flight about 14 days after hatching.

A somber low pitched cooing, "*Who-ah, whoo-whoo-who*" (the last three notes lower than the first), repeated from four to six minutes long makes a "joy in one" as they expound this melodious call. Some suggest this is a sad song of hopeless sorrow, while others would suggest the call is of tender love and devotion; the listener's mood, perhaps, is responsible for the determination.

So many seem to call asking for help in identifying this delightful, slender, peaceful, and well-mannered pinkish bird with the sweet low voice. "Well," I answer, "They are Morning Doves! Their name refers to the sweetness they bring to each day." "But I hear their beautiful call in the evening, as well," my friend would reply. "Yes, they call in the evening, and you have as well. Thanks for sharing your experience!" We should all share pleasure, shouldn't we? However, to salute the world each day with the sweet melodies of the Morning Dove is, indeed, a gift of expression we all can enjoy. "*Who-ah, whoo-whoo-who.*" Their music truly lifts our prospective.

FINCH, HOUSE

This is a very gregarious little bird about 5 to 6″ long or "House Sparrow" size. House Finches are often seen in flocks of twenty to thirty birds. A small bill, short legs, and mildly notched tail distinguish it from other finches.

Its song is a sweet "*cheep"* in flight, especially when excited during takeoff. When perched, its extensive warbling and rhythmic notes end with a "*where"* or "*cheuurr"*.

Adult male's forehead, broad eyebrow, chin, nap, throat, breast, and rump are red to rose-red or orange to orange-yellow and streaked buff-gray beneath. Wings are dull brown and coverts are edged with buff to dirty-white. The female's head, back, primaries, and tail are brown to dark-brown. The breast, belly, and flanks are striped with buff and light-brown.

They feed on a variety of food: weed seeds, buds, blossoms, fruit, insects, and bread crumbs. As we observe activity at our bird feeders during winter months, it is easy to see bird niches in feeding. House Finch, House Sparrow (quite timid before the Finch), and occasionally even Red-shafted Flickers will eat directly from the bird feeder, while Dark-eyed Junko (Oregon race) will eat from the porch beneath the feeder. Bird feeders attract birds, and these feathered creatures (while eating), invite danger or predators.

Native to the semi-arid lands of western North America and Mexico, they are found from southern Canada south to Mexico, and inland from the west to Texas, Colorado, and Montana. In 1940 some House Finches were found for sale in Brooklyn, New York, as Hollywood Finches. A Dr. Edward Fleisher knew it was illegal, so the pet shops apparently freed their stock of House Finches to avoid prosecution, thus beginning a race of Eastern birds. They now range from New England south to Florida and from Michigan east to Mississippi and Tennessee.

Two to five bluish-white lightly streaked or spotted eggs are laid. The House Finch makes a tightly woven compact nest in a woodpecker hole, tree cavity, building, or dense foliage, or it may even be attracted to a nest box.

The House Finch's need to sustain its high metabolism through long winter nights and short winter days demands a constant search for food. Although there is a modest reduction at night for fuel, this species doesn't approach torpid, thus making the daily acquisition of fuel to keep their little furnaces going full blast each night is very real. As one of the flocks finds food, that success of the find always brings a cheer of chirpings, songs, and frolic!

We seem to enjoy watching the House Finch most during the holiday season (Thanksgiving, Christmas, Hanukah, Kwanzaa, and New Year's Day). Picture this sighting of fresh new powdery snow, the bird feeder full of black-striped sunflower, oil sunflower, and Nyjer thistle seeds. As we see family, friends, and grandchildren looking on, we love these special occasions. In fact, they enjoy watching the birds feed almost as much as they enjoy feeding themselves. As we stock up on bird seed and food for these special times, let us remember those who yearn for such an occasion and help them to find an enjoyable holiday meal. When we share of our substance, we cause joy; then is not our happiness made more complete?

BLUEBIRD, EASTERN

The Eastern Bluebird is medium-sized, 6 to 7″ long. The size is important in its identification, for many birds are blue: Blue Jay, Indigo Bunting, Blue Grosbeak, and the Lazuli Bunting at first glance can be confused.

This lovely bird is seldom seen now but once was as commonly known as were robins. House Sparrows and Starlings, both alien introductions, claim homes they cannot make for themselves. It seems man must nest box the world to save the Eastern, Western, and Mountain Bluebirds.

A melodious rich whistling and warbling "*chur-lee*" is rendered for 20 to 30 minutes at a time. The common call is just a rich "*chur-lee*". It is an unforgettable low and attractive song.

The male has a bright blue back, wings, tail, and rust colored neck, breast, and flanks. The belly is white. The female and juvenile are smaller, 5½ to 6″, and duller; the young's colors are spotted. The female and juvenile each have blue in their wings and tail. The breast is rust; head and back are gray-blue; and all three have white eye rings.

They perch in a bent vertical style and feed mostly as they make little forages to the land. They eat insects, spiders, and berries and will pause at feeders for some peanut butter.

The nest is a cup of grass and stems 3 to 20′ above the ground in a nest box, woodpecker hole, or other natural hollow space.

Eastern Bluebirds are found in open woodlands, orchards, and farmlands. They are the only bluebird east of the Great Plains. But it is also found in central Northern Mexico and Texas, as well.

My friend, Virgil, told a story: *One day, he had finished eating something on his back porch and was carrying the empty bowl back to the kitchen from the porch. He remembered how the bowl was orange in color. While he walked, a little wild bluebird flew up to him and landed on the rim of the bowl. This stopped Virgil from walking, and the bluebird just sat there on the edge. After a few minutes, the bird flew up a few feet in front of him and it appeared the bird would fly away; but instead, the bird turned and landed on Virgil's left shoulder. Virgil stood there for what seemed to him a long time, and then he decided to walk the length of the porch, and sit down on the porch steps. Virgil thought surely when he started to walk that the bird would now fly away, but even after he sat on the steps, the bird stayed on his shoulder.*

Virgil sat on the steps for many minutes, and then he got up and walked back across the porch and entered the kitchen doorway. Upon entering an enclosed room, the bird finally flew off his shoulder and started flying around the kitchen. Virgil then propped the screen door open and left the room so that the bird might find its way out and not be hurt. Later - the bluebird had flown out into the woods.

This story was sent to Patsy and me with a picture and a request to identify this bird. It was clearly an Eastern Bluebird. Perhaps this Eastern Bluebird just felt Virgil and Martha had a home for him. Beauty or loveliness came into their lives and brightened their day. Isn't that a neat experience?

CHICKADEE, BLACK-CAPPED

The Black-capped Chickadee is 4¾ to 5½″ long. It is a small short-billed, plump and active year-round song bird. This chickadee seems more visible when it comes to brightens winter days. This Black-capped chickadee is very friendly (no fear of man) and loves bird-feeders and backyard nest-boxes. With a little patience on our part, they can learn to eat nuts or seeds from one's hand.

Its cap and bib are black, and the cheeks are white. The back and shoulders are olive-gray. The lower back and upper tail feathers are buff-gray; breast, belly, and under parts are white, but the flanks are buff. Wings, feathers, and secondary are edged with white.

Its voice is broad and lower than the other chickadees, "*Chick-a-dee-dee-dee".* The Black-capped Chickadee uses a call in spring and fall, which varies from the *Chick-a-dee-dee-dee"* song. This call is a whistled, "*fee-dee-ee,"* and the first note is quite high. On the other hand, while incubation is being tended, they have been known to spit or blow air, which creates a whistling sound at an intruder.

This bird is most widespread through mixed and deciduous woodlands, clearings, and suburban areas. They are found all across the northern half of the states, in all Canadian provinces and north to just above mid-Alaska.

This bird is one my mother much enjoyed. We had corner 4 x 6′ windows to gaze through while eating at the kitchen table. Black-cap was a common sight. Perhaps it was because of their teamwork and cooperation in searching every limb and branch for food. Perhaps it was the agility or acrobatics used in their search.

The chickadees seem to act like work is never done, exemplified by pairs during nesting season. Their labors festively continue in fall, winter, and spring as they flock together in swarms for food. Their voracious appetites demand an acrobatic turn upside-down hunt for morsels, while their hummingbird-like metabolisms seem never satisfied with what has been found. A chickadee can change direction in 3/100ths of a second and can hold a sunflower seed with one foot like a hand, hang onto the bird feeder with the other foot, and shell the seed with its small beak. From flight, he can alight on a perpendicular tree trunk, climb to a nearby limb and circle it time and time again. They eat caterpillars, weevils, scales, spiders, larvae, seeds and berries.

They like to nest in natural tree holes after meticulously cleaning the cavity before building a home of mosses, plant fiber, and feathers. Nest boxes must be cleaned for the chickadee to call it a home.

My grandson, Sam, once ask me to come to his school and be his V.I.P (Very Important Person). I was to share with his class something we like to do, Bald Eagle watching. He asked me to bring some of my paintings of birds and share with his class friends our love of birds. I brought about six paintings. I chose birds commonly sighted in winter in our area. I would hold a bird painting up and ask the class, "Who knows this bird's name?" I was delighted when over one third of the class knew the Black-capped Chickadee. It seemed every student wanted to talk about this small, but well-liked bird. Over a dozen hands were up and had to tell a story about a viewing - what fun!

IBIS, WHITE-FACED

The White-faced Ibis is 19 to 26″ long. This medium sized heron-like bird has a long and downward curved bill; its legs and neck are quite long. The White-faced Ibis is basically a western bird and is very similar to the Glossy Ibis of the east. It gives a low *quacking* call in flight. In breeding season or during summer, a bold white line appears around its bill and eyes. In the remaining months of each year, the White-faced and Glossy Ibis are virtually impossible to distinguish from each other.

This is a dark glossy-bronze or chestnut-brown colored bird. The gloss reflects brownish-purple from above. The head, neck, and tail have rich purplish-chestnut tones. The wing coverts are a lighter bronze. The primaries are whitish-green, the base turns to violet-green, and the tips become deep brownish-red with hues of purple. The short white feathers border the rich maroon faced skin. The eyes and base of the bill are red. Its bill is pinkish-gray and its legs are a brighter pinkish-gray.

They are very companionable birds at all times. To see a single bird hooking the shallow water alone makes one believe his spot for finding food is so rich that when the flock arose in flight, he just could not pull away from what he had found. Often the feeding grounds are long distances from the nest roosts. They are tireless workers in their quest for food. With a down-curved bill out front and the neck extended, the wings beat quickly. It then sails with its long legs dangling out back. They form in V's or diagonal lines and appear dark and eerie against beautiful sunset skies.

They like large marshes and wet farmlands. In the north, they are found in British Columbia, Utah, southern Idaho, Wyoming and Nebraska. They are also found in abundance in tropical zones, such as Oregon, Arizona, Louisiana, Texas, Florida, California, Mexico and southern South America. They nest in colonies. The nests consist of reeds and sticks in a well-shaped cup attached to living stalks of reeds. They lay three to four blue-green eggs. Incubation takes about three weeks by both sexes. The young leave the nest about four weeks after hatching. Their youth, like our youngsters, go right for the mud and puddles, thinking it is fun.

They eat crawfish, water insects, and frogs. The flock thinks nothing of flying ten or twenty miles to find food, then spending hours probing to obtain a meal before returning to roost or nest.

The slow-witted White-faced Ibis seems to appear like a witch or a huge wasp-like bird when covering the twilight sky in the fall, especially near Halloween. Their niche in nature, defined by the food they eat, has indeed given them some unusual peculiarities. How would you like to have a long hooked nose designed to sense food in a bog and spend most of your life in one quagmire or another, repeatedly trying to bag one small morsel then an additional worm, etc. out of the mucky mire? When we watch them work, we see that they are unyielding in their prying. Their down-turned bills are surprisingly sensitive to touch and taste, which indeed serves them well. These Ibises fly with soul mates, work alongside of friends, and build their homes near family. Close inspection reveals their colors, hues, and shine to be quite stunning. Indeed, we might need to change our minds about having both feet and nose in the mud because, in fact, they are quite colorful and happy, it seems!

GULL, WESTERN

This stately gull was captured on film while my wife and I were viewing the beautiful scenes on the "Seventeen Mile Drive" in the Monterey, California area. It was painted afterwards. This is a bulky or large gull and is 22 to 26″ long. It is the only widespread great, dark-mantled, thick billed gull on the west coast. The adult has an immaculate white head, neck, and breast, with a dark, slate blue-gray mantle which merges into black at the wing tips. Distinct white spots show in the black near the tip, as well. The bill is a massive bulbous size. It is yellow with a red spot near the tip on the lower mandible. The legs and webbed feet are pink. Southern birds have a very dark, slate-gray mantle, but in northern or central California, the mantle lightens to a medium gray; yet it is still slightly darker than the California Gull. In winter, the head shows streaks of brown. First year: the head and breast are gray-brown speckled with a blackish bill. Primaries are black, the rump is mottled white, and the legs and feet are dull pink. Winter plumage is distinctly different. The third winter bird has some quite different changes. The bill, now a dirty pink color, has black spots near the tip of both mandibles, somewhat resembling a Ring-billed Gull. The legs and feet are dull pink, the head and breast have a dusky mottling, and the tail shows a dark black band in flight.

It is found along the Pacific Coast from Washington, south to the Gulf of Baja, California. The glaucous winged gull, which resembles the Western, claims the coast to the north, and a Yellow-footed Gull, which is also easily confused with the Western Gull, which claims Mexican coasts. The nature of these gulls keeps them along the coasts, for they rarely ventures very far inland. They are really coastal birds.

Fish is their main sustenance, but they eat shrimp, garbage, and eggs and chicks of other nesting sea birds. Because of their size, they are quite successful in feeding out of others' nests. The male is the main provider; the female assumes almost all brooding responsibilities.

It breeds in colonies on offshore islands along pacific coasts. Some of the islands are now protected for sea bird colonies. Nests are made of grasses, seaweed, depressions, or rocks. Three buff-olive-brown eggs with blotches are laid. They hatch in 25 to 29 days. Young find flight in 40 to 45 days. The male defends the nest and territory, while the female broods the eggs or chicks.

It most commonly utters a low pitched series of "*cuk, cuk, cuk*" and also gives a loud squealing call.

This great bird appears as though he is standing at attention in parade attire: alert, and vigilant, with every part of his decorum clean and ready for inspection with his eyes watching the sea, guarding our pacific coast with honor and vigilance.

Perhaps each time we visit our west coast we might remember those who have defended our freedom. Many young men and woman have endured the regiment of military training, thus graduating as units of purpose to provide for our defense. Each has been a gleaming example of "spit and polished" training. Although these are visionary dreams, this great gull can remind us of those who stand on guard to protect our precious liberties each time we enjoy the sighting of this strong bird. The Western Gull's large size and stately perch seems always watchful: a lookout on guard. May our armed services always keep us free!

OWL, SNOWY

The Snowy Owl is a large arctic owl with a striking round head (no ear tufts). This snowy raptor has thick heavy plumage which completely clothes this bird from head to toe. Its lace feathers cover the toes and talons. The Snowy has a body length of 20 to 27″ and a wing span of 4½ to 5′. The raptor's stile is mostly hidden. The distinct gleaming yellow eyes appear to be looking from behind a mask. The adult male is almost pure milk white with very few bars or spots. The female has more spots than the male, and the immature has yet more bars and speckles than either of the adults. The eyes have no movement in the socket and are set forward in the round globe face. They find direction by the neck muscles which provide horizontal and vertical access; thus giving sight in all directions.

The extreme differences in length of day cause this arctic owl to hunt both day and night. To make bright daylight pursuit possible, this raptor possesses excellent hearing, night vision, silent soft flight, and eye protection (the colored irises expand as the pupils close down). The ears themselves supplement sight. They are uneven (one is higher on the head than the other), and are very sensitive, causing range identification. Like most owls, they would prefer to hunt after dark; obviously, however, this bird's pursuit to live must be performed in daylight.

They can be found nesting in the high latitudes of the tundra. Their main source of food is the Arctic Lemming, but they will eat birds (manly waterfowl), mammals, fish, and carrion. Their powers to pursue include the killing of an arctic hare or the overtaking of a grouse while in full flight. In winter, when lemming production declines (every 3 to 5 years), the bird is commonly sighted much further south in the northern states, especially along the Atlantic Ocean and around the Great Lakes. The author's first sighting of this great white bird was in northern Michigan's Huron Mountains; thus confirming their southern pursuit during lean lemming years. Snowy will kill and cache carcasses in the autumn, which they use to help them survive the short deep cold winter days. However, they will eventually go south. The variety of birds and mammals obviously expands as they leave the north in search for food. Constant provisions are necessary to maintain body heat in their chosen environment.

They find dry ground depressions usually on a small hill in the tundra, and then they line the nests with grass, feathers, and moss. They normally lay five to seven white or cream colored eggs, which are oval or ellipsoidal in shape. The incubation (by female) is about thirty-four days, and flight happens fourteen to twenty days later.

This owl is mostly silent. However, when it does call, it is a single "*hoot,*" but during breeding it is reported to give a "*croak,*" growl, or whistle sound.

Perhaps we could ask why this bird was given liquid spying yellow eyes which will not rotate or for what reason one ear is lower than the other. Differing from other owls, his eyes are equipped to follow after prey day or night. These differences deserve qualified ornithologists to answer, but for me as an artist, background is what counts. View a deep dark green conifer forest in winter with this heavenly white bird perched; or perhaps a better vision might be of a clear dark-blue sky background with a strong and large white round headed (angel-like) owl in soaring flight. However, I chose a glimpse of the Northern Lights (Aurora Borealis) in the above rendition. Whatever surroundings are chosen, this white bird is strikingly handsome.

Index

Made in United States
North Haven, CT
06 November 2026

21797415R00071